ADVANCE PRAISE

"Reading *From Chalkboards to ChatGPT* as a retired educator fills me with excitement and a bit of envy. The innovative approaches to teaching and the potential for AI to revolutionize the classroom experience are astounding. This book is a treasure trove of insights that I wish were available during my teaching years. It's a must-read for current educators eager to embrace the future of teaching."

Robin Shaffer,
35-year High School English Teacher and Department Chair

"This book brilliantly demystifies AI in education. It's not just about the tools; it's about transforming the learning experience. The balance between ethical considerations and practical applications is perfect for any educator looking to make a meaningful impact in their classroom."

Dr. Tony Thacker,
Former Assistant Superintendent for the Alabama State Department of Education

FROM CHALKBOARDS TO CHATGPT

Harnessing the Power of AI
in the Classroom

BY MICHAEL MARCHIONDA

From Chalkboards to ChatGPT

www.wisdomk12.ai
ISBN: 979-8-218-34524-2
First published in the United States in 2023 by Wisdom Media Group.

Cover Design by Chris Bannon
Layout by Buzzhoney
Printed in the United States of America

CONTENTS

DEDICATION

This book is a tribute to the incredible people who have illuminated my life's journey with wisdom, love, and inspiration. They have left an indelible mark on my heart, and their influence has been the driving force behind my pursuit of knowledge and the written word.

To my cherished daughters, Morgan and Reagan, you are the brilliant stars in my sky, constantly encouraging me to explore the boundless possibilities of the future. Your curiosity and enthusiasm remind me to embrace each new day with open arms.

To my dearest mother, Kathleen Marchionda, you are the original storyteller in my life. Your passion for reading and writing has ignited the creative spark within me, and I am forever grateful for the foundation you've laid.

To my beloved grandmother, Mary Ellen Verner, your spirit is a testament to the enduring power of learning. At 92, you embraced the world of technology, Kindle in hand, teaching me that age is no obstacle to growth and discovery.

To my treasured colleague, Ruthie Robbins, our journey together has been nothing short of transformative. From the first day in your General Business class to our collaborative work today, your mentorship and dedication have been a guiding light in my life. Your pioneering work in the realm of AI education continues to inspire me.

With profound gratitude, I dedicate this book to each of you. Your lessons, love, and unwavering support have shaped me into the person and writer I am today.

Introduction: Magically Empower Your Teaching

There are some people you meet in life whose influence becomes immeasurable, shaping the trajectory of your professional and personal journey. For me, Ruthie Robbins is that person. Our story began in a 9th-grade Intro to Business class. She was the educator who brought authenticity to every task, and I was the eager student, hanging on to her every word. We designed and sold t-shirts that arrived two weeks late, on the morning of the last day of school. When a few well-meaning students made mistakes in filling the orders, we were out of time to correct them, and the project went awry. As a student accustomed to angry teachers, I remember being surprised that Ruthie was undaunted, telling us we sometimes learn more from our mistakes than from our successes.

A few years later, our paths intertwined again, but this time I was Ruthie's student teacher for English 7 during my senior year in college. From teacher-student to colleague to my final role in the district as Director of Curriculum, our bond grew stronger, rooted deeply in our shared passion for education.

Yet, through all our collaboration, one image of Ruthie remains etched in my mind. It was a photo that was somewhat amusing but truly sad to see. It was a poignant photo that her family took when they woke in the morning to find Ruthie asleep on the living room floor, still wearing the clothes she wore to school the previous day, surrounded by stacks of student essays. Although Ruthie often said she was "up all night" to give timely feedback and keep the writing momentum going, I had not realized how literally she meant that. The photograph represented more than just a fleeting moment of fatigue; it symbolized the struggle of countless educators who are bogged down by teacher overload, with too many papers to score while struggling to create lessons that are not only pertinent to the student's individual needs but also interesting and aligned to state standards.

This scene, however, isn't where Ruthie's story ends, nor where ours begins. From that pivotal moment, she and I embarked on a transformative journey through the realm of Artificial Intelligence, seeking answers to the age-old conundrums of teaching: How can we shed the mundane tasks without losing the magic of genuine learning? How can we connect more deeply with our students when managerial tasks constantly vie for our attention?

We discussed that media reports seemed to be popping up everywhere about teachers, young and old, leaving the profession in unprecedented numbers. Articles with titles such as "I Just Found Myself Struggling to Keep Up: Number of Teachers Quitting Hits New High" (*USA Today/Chalkbeat*, March 2023); "More than half the teachers are looking for the exits," (*NPR*, February 2022); and "Teachers, Facing Increasing Levels of Stress, Are Burned Out" (*NY Times*, March 2023) have become common, with clear data from multiple states as evidence for the unfortunate trend.

In "The Teachers Who Aren't Coming Back This Year," (*Chalkbeat*, September 2022), eighth-year social studies teacher Abbey Nova responded to the question "What do you think is the biggest challenge facing teachers today?" Her answer typifies that of dozens of teachers to whom we have recently spoken.

> The amount of labor involved, physically, mentally, spiritually, is very intense.
>
> The amount of juggling required to get through a day is truly exhausting. This was the first time that I went on a spring break and didn't feel completely spent. We need to slow down. Teachers need to have more time to create curriculum that is thoughtful, liberating, and meaningful for students.

We wonder if Abbey might have stayed if she were able to harness the power of AI to streamline her own tasks and enhance her students' educational journey. Come back, Abbey.

In this book, you won't just read about Ruthie's transformation from overwhelmed to overjoyed. You will be handed the very blueprint she used to reclaim her teaching profession and personal life. But here's the game-changer: By investing in this book, you are securing for yourself actionable, specific examples and lessons that you can start implementing tomorrow. We're not talking about theoretical fluff or distant future tech. We're presenting tangible tools and strategies, real lessons and projects honed by Ruthie's own experiences, which will allow you to harness the power of AI immediately.

If you've ever wished for a magic wand to reduce your workload and reignite your passion for teaching, consider this book your spellbook. Within these pages lies Ruthie's transformative journey, and by walking in her footsteps, you too can experience a renaissance in your teaching approach. Ready to embark on the journey to reclaim your time, passion, and joy in teaching? Jump in, and let's revolutionize education together.

TIME SAVERS & REFLECTION MOMENTS

As you embark on this journey through the pages of this book, keep an eye out for the ***TIME SAVERS*** *and* ***REFLECTION MOMENTS*** *peppered throughout each chapter. These segments are specially designed to offer you immediate, actionable strategies to transform your teaching and moments to pause, reflect, and visualize the impact of integrating AI and ChatGPT into your classroom. Think of them as your personal beacons, illuminating the most direct path to a future-proof, engaging, and efficient teaching experience. Jump in, and let's innovate together!*

CHAPTER 1

THE EVOLUTION OF EDTECH—EMBRACING AI AND CHATGPT

In every generation, classrooms evolve. The chalky blackboards of the past gave way to whiteboards, which in turn stepped aside for interactive digital screens. Each shift was met with its own mix of excitement, resistance, nostalgia, and humor. As teachers, it's amusing to reflect on how we've navigated these changes, from fumbling with the buttons of an overhead projector to mastering the art of Zoom backgrounds. This chapter journeys through the history of educational technology, leading up to the latest promising addition: AI and ChatGPT. Let's embark on this trip down memory lane, recalling those charming quirks of past tech tools, and then looking ahead at the possibilities our latest companions present.

A STROLL DOWN MEMORY LANE: EARLY EDTECH INNOVATIONS

Overhead Projectors: Remember when transparency sheets were the pinnacle of presenting information? The PowerPoint of yesteryear, the vaunted overhead projector, that hulking beast in the corner of the room, was once the crown jewel of technology in the classroom. Teachers would painstakingly write out notes, sometimes in color (if we were feeling fancy), ensuring that every student could see the material. And who could forget the art of carefully placing and replacing sheets without blocking the projection or creating a shadow puppet show? Then, of course, there were those daring moments when a teacher would attempt to write on the projector in real-time, resulting in many a backward-written word. Students took immense pride in being selected to erase old notes—an art of ensuring the ink didn't smudge everywhere. And not every class was equipped with its own overhead projector, so you had to sign up and hope you got the day you hoped for and that the lamp in the projector wouldn't burn out. I won't even go into the opaque projector. It looked like a modern-day MRI machine. Only the courageous used it!! Ruthie and I, sad to say, passed on the iron lung of projectors.

VHS & TV Carts: Ah, the TV cart! Its arrival in the classroom often incited whispers of excitement among students. A break from the norm, perhaps a documentary or a movie was on the agenda. For teachers, it was a mixed bag. Wrestling with the cords, ensuring the TV was plugged in and the VHS player was working, and then the ultimate challenge: finding the correct rewind spot on a well-worn tape. We all shared the mutual agony of static interference and the inevitable groans when the teacher had to adjust the tracking dial. And, of course, the ritual of "Be Kind, Please Rewind." Not to mention that when the lights went out, so did the students' attention. I can still remember during my student teaching with Ruthie as my cooperating teacher how excited I was to integrate video into the lessons. I kept the lights on, presented small snippets, and thought I was on to something. Until . . .

Early PCs: The introduction of personal computers in classrooms was groundbreaking. Students would line up, waiting for their chance to explore the digital realm. Programs like Microsoft Paint turned everyone into an artist, while games like Oregon Trail made computer time the most anticipated part of the day. Floppy disks were the magical devices that held our data, though many an assignment was tragically lost to their unpredictable nature. Teachers often embarked on a steep learning curve, from mastering basic operations to troubleshooting the frequent hiccups. There was always that one computer with a rebellious streak, testing a teacher's patience daily. Again, I remember how excited I was as a first-year rookie teacher to get the huge boxes filled with a clunky monitor that seemed to weigh 100 pounds that would cover my entire desk and the desktop computer that I always kept kicking with my feet under my desk. It helped me do things that before were all manual and time-consuming: take attendance, create lesson plans, and record grades in an electronic gradebook that actually did the calculations for me.

Interactive Whiteboards

From video to computers, technology for all teachers was evolving rapidly. Smart Boards were a marvel when they first graced our classrooms. Then suddenly, screens could respond to touch, and dynamic presentations were possible. Lessons became more interactive and visually engaging. But, like any tech, it had its quirks. Calibration became the dreaded start-of-day routine for some teachers, and many found themselves unintentionally drawing lines when all they wanted to do was point at something. There was the usual trial and error–figuring out which pens worked, the art of not losing the pen cap, and the discovery that, yes, you can accidentally erase your entire presentation with an errant touch.

The heartwarming part? For every technological hiccup, there was a learning moment, not just about the tech itself, but also about patience, adaptability, and the joy of shared discovery. As we delve deeper into this chapter, you'll find that while the tools have evolved, the essence of teaching—and the amusing challenges we face—remain beautifully constant.

THE DAWN OF THE DIGITAL AGE: MODERN TOOLS AND THEIR QUIRKS

PowerPoint and Slides

From overhead transparencies, we transitioned to the digital world of presentations. PowerPoint and similar tools made every teacher a designer, for better or worse. We ventured into a world of fly-in transitions, WordArt, and those dreaded sound effects that we eventually realized weren't necessary for every slide. And who among us hasn't suffered the mini heart-attack of a misbehaving slide or the panic when you can't find that one essential slide that disappeared mysteriously? The most adventurous even dipped their toes into custom animations, leading to slides where content danced, twirled, and occasionally, caused a bit of dizziness. And by the way, what is the WingDings font all about? Lol.

Learning Management Systems (LMS)

Platforms like Moodle, Blackboard, and Google Classroom promised a new era of streamlined education. Assignments, resources, grades—all in one place! And while they delivered on many counts, they also introduced us to a slew of new challenges. "I uploaded my homework, I swear!" became the new dog-ate-my-homework excuse. Password forgetfulness reached epidemic proportions, and the phrase "I can't find it" echoed endlessly in classrooms worldwide.

Tablets & Smartphones

Bring Your Own Device (BYOD) policies heralded the integration of personal tech into learning. Educational apps, eBooks, and interactive simulations held the potential to revolutionize teaching. However, this new era also ushered in distractions galore. Teachers became adept at identifying the classic signs: the student gazing intently into their lap, the suspiciously angled book hiding a phone, or the headphones sneakily threaded up through a hoodie. And yet, amidst these challenges, we also discovered new, effective ways to make learning relevant in a digitally connected age.

THE AI & CHATGPT REVOLUTION: MORE THAN JUST A TOOL

With every advancement in educational technology, teachers have adapted, molded, and sometimes wrestled with the tools at their disposal to craft meaningful learning experiences. Now, as we're introduced to ChatGPT and other AI tools, we're not just looking at another tech addition, but rather partnering with a game-changing ally.

What is ChatGPT?

At its core, ChatGPT is a conversational AI developed by OpenAI, but think of it as your super-powered teaching assistant, always available, and incredibly well-informed. It can generate content, answer questions, provide feedback, and even assist in lesson planning. With a vast amount of information at its fingertips, it's like having an instant expert ready to help.

While its efficiency and capability can help reduce the administrative burden, ChatGPT is more than just automation. It's a tool that can adapt, learn, and provide personalized support, making the teaching experience richer and more interactive. Imagine instant feedback on student essays, brainstorming ideas for a class project, or even getting suggestions for classroom activities—all with ChatGPT.

Just like any new tool, there's a learning curve with ChatGPT. And yes, there will be quirks, unexpected responses, and moments of amusement (or bemusement). But, as history has shown, teachers excel at taking new tools, early hiccups and all, understanding them, and then shaping them to fit the unique needs of their classrooms. ChatGPT and AI are no exceptions.

With each chapter of this technological story, teachers have emerged as the true heroes—adapting, innovating, and always prioritizing their students' needs. As we embark on this new chapter with ChatGPT, let's embrace it with the same spirit, curiosity, and passion. After all, the future of education is always a blend of the past's wisdom and the possibilities of tomorrow.

AI IN POP CULTURE: FROM SILVER SCREENS TO CLASSROOMS

Pop culture has long been captivated by the potential, and *often the peril, of artificial intelligence. Sci-fi enthusiasts* among us might remember HAL from *2001: A Space Odyssey* or the self-aware Skynet from *Terminator*. And who could forget the slick and somewhat sinister robots from Will Smith's *I, Robot*? But for every ominous cinematic AI, there's a lovable WALL-E or the fiercely protective TARS from *Interstellar*. So, where does ChatGPT fit in this spectrum?

Before we jump in, let's remember that Hollywood thrives on drama. Artificial intelligence generating efficient lesson plans or suggesting reading materials might not make for a blockbuster movie, but it's precisely where ChatGPT shines in the real world. It's neither a world-dominating evil nor a quirky metal sidekick; it's a tool. A smart one, yes, but still a tool.

Instead of being overpowered or overshadowed by AI, teachers stand as the actual directors of this show. While Will Smith might have needed to save humanity from robots, our teachers only need to save a lesson plan from becoming monotonous. And with ChatGPT, they have a co-star ready to assist.

Of course, there'll be moments when ChatGPT might give a

response worthy of a sci-fi comedy spin-off. But these quirks can become learning moments, laughter-filled class memories, or just another teaching tale to share in the staff room.

THE WHY BEFORE THE HOW: THE PURPOSE BEHIND INTEGRATING AI

Before diving headfirst into the vast sea of AI's potential, it is essential to reflect on the "why." Just because a tool can do something doesn't always mean it should. Understanding the motivation behind integrating ChatGPT can make its inclusion more purposeful and impactful.

Every student is unique. With varied learning speeds, styles, and strengths, a one-size-fits-all approach often falls short. AI can help tailor learning experiences, offering resources and tasks suited to individual needs.

Reclaiming Time

Admin tasks, grading, lesson planning—these necessary tasks consume vast amounts of a teacher's time. What if a portion of that could be reclaimed? ChatGPT can handle many administrative chores, allowing educators to invest more time where it truly counts—with their students.

Bridging Knowledge Gaps

No teacher can know everything (although, to their students, it often seems like they do). With ChatGPT, educators have a backup. A quick query can fill in knowledge gaps, provide up-to-date information, or even suggest new teaching methodologies.

Just as chalkboards made way for digital screens and encyclopedias were replaced by online databases, AI is the next step in educational evolution. By integrating it effectively, teachers ensure they remain at the forefront of delivering contemporary, relevant, and engaging education.

As we continue this journey, it is evident that AI, with tools like ChatGPT, presents an expansive realm of possibilities. But it's the heart, intuition, and passion of a teacher that will truly define the future of education. Whether it's navigating the quirks of new technology, laughing along with pop culture references, or critically examining the reasons behind our choices, educators remain the architects of our learning landscape.

YOUR FIRST STEP TO A BRIGHTER, AI-ENHANCED FUTURE

By now, you've journeyed through the evolution of tech tools in the classroom, chuckled at some nostalgic and cinematic AI blunders, and envisioned the potential that ChatGPT and AI bring to your educational toolkit. But what's next? How do you move from understanding to action, from curiosity to capability?

Start simply. Maybe it's a quick chat with ChatGPT to help brainstorm your next lesson plan topic, or perhaps it's asking for background and anecdotes about the author of the book you're introducing next week. Small successes build confidence.

Reflection: *How do you typically approach new technology or methods in your teaching? What is a small, achievable goal you can set for yourself with ChatGPT this week?*

As you begin, not everything will be smooth sailing. There might be odd answers or unexpected responses from ChatGPT. Embrace them. Laugh, learn, and remember—it's these unique experiences that often leave lasting memories and teachable moments. I can't explain how much fun Ruthie and I have had playing with the tool and learning all along the way.

Reflection: *Think of a time you faced an unexpected hiccup in your teaching methods. How did you turn it into a learning experience?*

Connect with Colleagues

Share your initial experiences, both highs and lows, with fellow educators. Collaborative troubleshooting, exchanging ideas, and shared laughter can make the process less daunting and much more enjoyable.

Reflection: *Who in your educator network would be excited to explore ChatGPT with you? How can you foster a community of sharing and growth?*

Just as you instill a love for learning in your students, nurture that same spirit within yourself. AI is ever-evolving, and the best way to stay ahead is by continually exploring, questioning, and experimenting. Don't worry, it doesn't take much to master what you need from ChatGPT. Then you will keep adding to your skills with the tool, and you will grow along with it.

Reflection: *What drives your passion for teaching? How can integrating AI like ChatGPT further fuel that passion?*

ACTION STEPS TO KICKSTART YOUR AI ADVENTURE:

1. **Join a Community:** There are numerous online communities where educators discuss the intersection of AI and education. This will give you a platform to ask

questions, share successes, and learn from others' experiences. See Appendix A for a list of online communities and platforms.

2. **Set a Goal:** Choose one lesson or task this week where you'll incorporate ChatGPT. It can be as simple as generating discussion questions or seeking clarification on a topic.
3. **Document Your Journey:** Keep a brief journal or digital document noting your experiences, questions, and revelations as you integrate ChatGPT. This will be invaluable as you progress.

To many, the inclusion of AI in their classroom is a daunting leap into the unknown. However, when viewed in light of all of the technological changes that have already been successfully implemented, AI is just one more iterative step in the effort to improve education and student learning. Remember, every technological evolution in the classroom started with educators like you taking the first step. The future is bright, engaging, and just a page turn away.

"AI HAS THE POWER TO CHANGE TEACHERS' LIVES FOR THE BETTER, FOREVER. THERE IS NO TIME LIKE THE PRESENT TO LEARN THESE SKILLS."

MIKE MARCHIONDA

CHAPTER 2

TAKING THE FIRST STEP: SETTING UP CHATGPT

The ambiance of Ruthie's workspace with scattered lesson plans, curriculum maps, and colorful sticky notes set the backdrop for our adventure into the world of ChatGPT. I was eager to show Ruthie the wonders of this tool and how it could amplify her teaching methods.

"Alright, Ruthie, let's get you plugged into ChatGPT," I began, navigating to the OpenAI website on her desktop. As the page loaded, I sensed a mixture of excitement and curiosity in her eyes.

Navigate to OpenAI.com

"First, we'll head to OpenAI's official website," I pointed out. Ruthie leaned in, absorbing every step.

ACCOUNT CREATION

"You'll need an account to get started," I explained. Ruthie quickly clicked on the "Sign Up" button and filled in her details. "Think of this account as your passport to a world of AI-driven educational magic," I quipped.

Once inside, I guided her cursor to ChatGPT's dashboard. "This, Ruthie, is where the fun begins."

Exploring the Features

COMPREHENSION MODE

"Let's start with Comprehension Mode," I said. "Try copying a paragraph from any article and pasting it here." Ruthie did just that and was left amazed at how ChatGPT succinctly summarized the content. "This will save me so much time when I need to distill lengthy texts for my class!" she exclaimed.

WRITING ASSISTANT

Next, I showed her the Writing Assistant. "Think of a sentence or phrase you use frequently in class." Ruthie typed in a generic feedback sentence she often used on student papers: "Compound sentences must be joined by a semi-colon or a comma followed by a conjunction." ChatGPT quickly offered ten alternative ways to express the same idea. "Wow, this can really help in diversifying my feedback," she noted with surprise.

Q&A MODE

"Lastly, the Q & A mode," I introduced. I prompted Ruthie to ask any question on a topic she'd been pondering. The answer ChatGPT generated was detailed and informative. "This could be a game-changer for preparing lessons and providing relevant background," she mused.

As we wrapped up, Ruthie leaned back, looking both surprised and relieved. "I honestly didn't think it would be this easy to sign up and get started. I always assumed it'd be way more technical," she admitted with a chuckle.

The straightforward process and the fact that the trial period was free added to the allure. Now armed with ChatGPT, Ruthie was poised to introduce a transformative touch to her teaching. "All teachers should know about this!" she proclaimed.

We both shared a moment of realization, eager to see where this new tool would lead.

CHAPTER 3

"TIME IN A BOTTLE: RECLAIMING YOUR TEACHING HOURS WITH CHATGPT & AI"

Every teacher has their list of those repetitive tasks that seem to eat up hours like minutes, tasks that are crucial but not always directly related to impacting a student's learning. Take grading, for instance. It's essential, of course, but the hours spent pouring over papers often feels like time that could be better spent engaging with students, designing innovative lessons, or simply recharging for the next day.

And then there's lesson planning. While crafting the perfect lesson is a joyful art, searching for resources, aligning them with standards, and ensuring differentiation can often become a time-consuming chore.

How about administrative duties? Attendance logs, communications with parents, scheduling . . . the list goes on.

Remember, as we transition through this chapter, we aim to make you see these tasks in a new light, unveiling ways that technology, particularly ChatGPT & AI, cannot only simplify them but also rejuvenate the joy in teaching.

Onwards, to a future where teachers get to do more of what they love to do and less of what they "must" do!

RUTHIE ROBBINS' GAME-CHANGING JOURNEY

"It's not just about the hours I've gained. It's about how those hours are spent. I've got more time for my students, for professional development, and even for my personal life!"

AI isn't about replacing. It's about enhancing. Ruthie's journey is a testament to what's possible when educators harness the power of AI, not to replace their unique skills and human touch, but to amplify their reach and impact. By taking back time and energy from perfunctory tasks, teachers can redirect their efforts to what truly matters: creating meaningful connections with students and inspiring a lifelong love for learning.

Ruthie's words summarize it best: "Using these tools doesn't make me any less of a teacher. In fact, it makes me a better one. I can focus on my students, innovate my lessons, and even find moments of self-care."

And as for me? Witnessing Ruthie's transformation from

a drained educator to a revitalized powerhouse was truly rewarding. I'm more convinced than ever that if we can provide teachers with the right tools and training, they can reclaim not just hours, but joy, passion, and the essence of teaching.

It's an exciting horizon. And with tools like ChatGPT and the ever-evolving landscape of AI, the dream of a balanced, enriched, and fulfilling teaching career is not just possible—it's within our grasp.

Reflection Moment: *Consider your current workload and the tasks that consume the most time. What new opportunities would open up for you if you could get some of that time back? How would your relationship with your students evolve?*

LESSON PLANNING IN A FLASH: THE DAY RUTHIE MET CHATGPT

During one of our brainstorming sessions, Ruthie vented about hitting a planning roadblock. "How about a dance with ChatGPT?" I suggested.

Ruthie, always curious, asked, "So, how do I start this . . . dance?"

I chuckled. "Just start by telling ChatGPT about your next topic, the age group, and any specific goals or objectives you have in mind."

Up for a challenge, Ruthie gave it a go. She hesitated a moment, then typed into the interface, *Planning a lesson for 8th graders on the themes of Harper Lee's To Kill a Mockingbird. I want to emphasize empathy, historical context, and critical thinking. Any ideas?"*

Ruthie had barely removed her finger from the "Enter"

button on the keyboard when ChatGPT responded. The suggestions spanned from analyzing character motivations to role-playing exercises, simulating the social environment of the era, and even recommending multimedia resources that offered a rich understanding of the period's history.

Ruthie's eyebrows shot up in surprise. "This is good. *Really* good!"

She continued her dialogue with ChatGPT, refining her lesson structure. *"How can I incorporate collaborative learning for the empathy theme?"*

ChatGPT suggested, *"Consider grouping the students and assigning them characters from the book. Each group can present a short skit, showcasing a pivotal moment from their character's perspective, emphasizing their emotions and challenges."*

By the end of her session, Ruthie had a detailed lesson plan, complete with activities, discussion questions, and even a list of supplementary resources. It wasn't just the time she saved; it was the fresh perspective and innovative ideas she gained that truly amazed her.

"The magic isn't just in the speed," she mused, looking at her new lesson plan. "It's in seeing teaching from a different angle. ChatGPT hasn't just given me a lesson plan; it's given me new ways to inspire my students."

And that's the moment I knew Ruthie truly grasped the power of this AI tool in her hands.

EMBRACING CHATGPT FOR EFFICIENT LESSON PLANNING

We've all been there: Sitting in front of our computer, staring blankly, and waiting for inspiration to strike up that perfect lesson plan. With the vast ocean of information available, sorting through resources can be overwhelming. What if you had a tool that could help streamline this process, tailor resources to your specific needs, and provide innovative ideas on demand?

Enter ChatGPT

Imagine having a conversation about your next lesson on the solar system. You express a desire to make it interactive and hands-on for your 6th graders. Within seconds, ChatGPT could provide a list of engaging activities, online simulations, and even trivia questions to pepper throughout your lesson. No more sifting through pages of Google search results; you have a personalized list ready to go.

TIME SAVER: Next time you're planning, pose a topic to ChatGPT and ask for three interactive activity ideas. You might be pleasantly surprised by the creativity of the suggestions!

AUTOMATING GRADING AND FEEDBACK

Feedback is a critical aspect of student growth; however, grading assignments, particularly repetitive ones, can become a tedious chore. What if there were a way to provide instant, constructive feedback without spending evenings buried under a pile of papers? What if I tell you there IS a way?

AI tools can assist with this. It's not about replacing the personal touch that teachers bring to feedback; it's about streamlining the process. For assignments that focus on factual accuracy (like math problems or grammar exercises), AI can quickly check and provide corrections. For more subjective assignments, AI can assist by highlighting potential areas of concern, allowing the teacher to focus on those sections, and providing your individualized touch.

Reflection Moment: *Think about a recent assignment you spent hours grading. How might AI tools have assisted in reducing that time? What other tasks or self-care activities could you have accomplished with those extra hours?*

As we delve deeper, our focus remains on harnessing the power of ChatGPT and AI to complement your skills and expertise, not replace them. These tools are here to elevate your teaching experience and provide the gift of time—a commodity every teacher wishes they had more of!

PERSONALIZED LEARNING WITH AI INSIGHTS

The beauty of education is that every student is unique, bringing a diverse set of strengths, weaknesses, and learning styles to the table. Traditional methods can make it challenging to cater to each student's specific needs. But what if you had a tool in your arsenal that could help?

With AI's data-driven insights, you can identify patterns in individual student performances, identifying where they excel and where they might need extra help. This isn't just about test scores; it's about understanding their interaction patterns, the types of questions they ask, and even their collaborative tendencies.

For instance, let's say you're teaching a lesson on Shakespeare. ChatGPT can help formulate comprehension questions tailored to various difficulty levels. Based on AI insights, you can assign these questions to students based on their current understanding, ensuring that everyone is challenged just the right amount.

TIME SAVER: *Next time you're preparing a lesson, ask ChatGPT for differentiated questions or tasks based on three levels of understanding: beginner, intermediate, and advanced. This can be a starting point for personalized instruction.*

ENHANCING STUDENT ENGAGEMENT WITH INTERACTIVE AI TOOLS

We've all faced the challenge of keeping our students engaged, especially with the increasing distractions of the digital age. AI and ChatGPT can be game changers in this realm.

Imagine incorporating an AI-driven chatbot into a lesson where students can ask questions and get instant, informative responses. For a history lesson on Ancient Egypt, students might interact with a "Pharaoh Chatbot" that provides first-person insights into the life and times of Ancient Egypt. The possibilities for interactive, engaging learning experiences are endless.

Reflection Moment: *Think about a topic that your students traditionally find challenging or dull. How might you use ChatGPT to create an engaging, interactive experience around that topic? How do you think your students would respond to such an innovative approach?*

ADMINISTRIVIA

Having worked with Ruthie for decades, I knew that forgetfulness was one of her personal challenges. She has a magnet on her file cabinet that reads, "Not to brag, but I can forget what I'm doing while I'm doing it." When she was twelve, her father made her a "reminder necklace" by attaching a small spiral notebook to a bicycle chain. Ruthie's forgetfulness went beyond typical forgot-where-I-parked-my-car anecdotes. She had countless entertaining stories that kept us in stitches, like when her the table for a carefully planned dinner party looked sparse. As guests arrived, she realized that she had never picked up the spiral cut ham or the bread and rolls from the bakery. The meal was unintentionally vegetarian. In another tale, she forgot to get someone in the family to button the back of her dress for church. This wouldn't have been so bad if she wasn't the organist or if the organ weren't placed at the front of the church, her back to the congregation. She only realized the faux pas when someone stepped out of their pew to tap her on the shoulder. She heard the muffled collective chuckle of the congregation when she left the organ bench to take care of it.

My most immediate suggestion was for Ruthie to explore Trello. "Think of it as a digital pin board, Ruthie. It's going to revolutionize how you manage your to-do list and classroom projects," I asserted.

At first, she was skeptical; however, upon integrating Trello with her existing systems, Ruthie quickly saw its potential. She set up individual boards for each class, tracking assignments, due dates, and student projects with ease. She even created a personal board for her administrative tasks, including parent communications, meetings, and professional development. The plugin feature allowed her to connect to her email, ensuring no missed communications or overlooked tasks.

But the real charm? Power-Ups. These are like mini plugins that can be added to a Trello board. One of Ruthie's favorites was the Calendar Power-Up. She had an integrated view of all her tasks, assignments, and meetings into one calendar, streamlining her planning process.

It didn't stop there. With the Butler Automation, she set automatic reminders for her tasks, ensuring that she never missed an assignment deadline or forgot about an upcoming parent-teacher meeting.

"I can't believe how much time this simple tool is saving me," Ruthie remarked one day, her enthusiasm evident. Not only did Trello bring structure to her administrative tasks, but it also freed up hours that she previously spent juggling multiple platforms and manual reminders.

ADDING UP THE HOURS: OUR EUREKA MOMENT

After a couple months of these tweaks and trials, Ruthie and I sat down, adding up the time saved. As we reviewed her weekly schedule, the numbers spoke loudly: She had more than an entire day back in her week!

BEYOND TIME: UNEARTHED BENEFITS

Ruthie realized, "It's not just about the hours I've gained. It's about how those hours are spent. I've got more time for my students, for my personal life, and even for professional development."

As time went on, Ruthie and I affirmed that it wasn't just about the hours. Our discussions often circled back to how her grading became precise, her lessons fresh and personalized to ability levels, and her students even more engaged and responsive.

AI isn't about replacing, it's about enhancing: Ruthie's journey is a testament to what is possible when educators harness the power of AI, not to replace their unique skills and human touch but to amplify their reach and impact. By taking back time and energy from perfunctory tasks, they can redirect their efforts to what truly matters: creating meaningful connections with students and inspiring a lifelong love for learning.

Ruthie's words summarize it best: "Using these tools doesn't make me any less of a teacher. In fact, it makes me a better one. I can focus on my students, innovate my lessons, and even find moments of self-care."

As for me? Witnessing Ruthie's transformation from a drained educator to a revitalized powerhouse was truly rewarding. I'm more convinced than ever that if we can provide teachers with the right tools and training, they can reclaim not just hours, but joy, passion, and the essence of teaching.

The Future Beckons: As we close this chapter, I invite you to look to the future. A future where technology works hand-in-hand with educators, creating an environment where students thrive and teachers flourish. It's an exciting horizon. And with tools like ChatGPT and the ever-evolving landscape of AI, the dream of a balanced, enriched, and fulfilling teaching career is not just possible—it's within our grasp.

Reflection Moment: *Consider your current workload and the tasks that consume the most time. How might integrating technology, particularly ChatGPT and AI, change your teaching dynamics? What new opportunities would open up for you? How would your relationship* *with your students evolve?*

Our journey is just beginning, but the possibilities are endless. Here's to rediscovering the joy in teaching and making every moment count!

CHAPTER 4

ENSURING ETHICAL USE IN OUR CLASSROOMS

Before embarking on the transformative journey of intertwining AI with our teaching fabric, it's imperative to set our bearings right. Just as a seasoned sailor checks the sturdiness of his ship and ensures the safety of his crew before setting sail, educators too must first ensure that the digital tools we weave into our pedagogy safeguard our most treasured assets: our students. This chapter is dedicated to that very essence, equipping educators with the compass and map to navigate the vast digital ocean safely, ensuring that every student embarks on a journey that is not just enlightening but also secure and fair.

PROTECTING THE MOST VALUABLE TREASURES (OUR STUDENTS)

Just as a museum protects its priceless artifacts with

state-of-the-art security systems, so too must we protect the valuable data of our students. Consider the data as each student's unique legacy: their growth, their interactions, and their feedback. Ruthie and I were once evaluating digital platforms and were drawn to ones that treated student data with the utmost respect. Platforms like ClassDojo and Google Classroom stand out as guardians of this legacy, ensuring our students' information remains protected from any external misuse. Navigating the vast landscape of digital tools can be daunting, especially when considering data privacy. As educators, it is crucial that we are vigilant. We have provided key pointers to help you select and use technology responsibly:

- **Transparent Privacy Policies:** Always go for tools and platforms that have clear, understandable privacy policies. If you can't understand a tool's policy or if it's hidden behind layers of complex jargon, it might be best to reconsider its use.
- **Data Encryption:** Ensure the tools you use encrypt data. This means transforming student data into a code to prevent unauthorized access. Tools like Google Classroom and ClassDojo, for instance, use encryption to protect user data.
- **Minimal Data Collection:** Opt for platforms that only ask for essential information. If a tool is asking for more data than it needs to function (like a student's home address when it's not required), it's a red flag.
- **Control over Data:** Platforms should allow educators and schools to control their data. This includes being able to review, modify, and even delete data.
- **Third-Party Sharing:** Be wary of platforms that share

data with third parties. Always check their sharing policies and whether they are giving data to advertisers or other entities.

- **Regular Updates and Patches:** Good digital tools will frequently update to patch any vulnerabilities. Ensure whatever you use is kept up to date.
- **Training and Awareness:** Continually educate yourself about the latest in data privacy. Platforms like the International Society for Technology in Education (ISTE) offer resources and training on this.

Reflection Moment: Review the tools you're currently using in your classroom. Do they align with the above pointers? If not, what steps can you take to ensure your students' data is protected?

Always check the data protection policies of any new platform you're considering. Your students' digital legacy deserves the gold standard of protection.

FAIRNESS FOR ALL

Ruthie felt that having the AI feedback so closely match her own helped with students' sensitivity to fairness. When some students receive a low score, they are quick to blame rather than looking at themselves first. It's not uncommon for a student to believe that a teacher favors other students or simply doesn't like them, despite the specificity of rubrics and teacher feedback. With AI scoring, however, students readily accept that the feedback is objective.

This brought up a pivotal discussion on fairness. Just like in that classroom, as we integrate AI into our teaching, it's essential to choose tools that ensure every student gets an equal chance to succeed. By vetting platforms for inclusivity and fairness, we're leveling the playing field.

In the realm of AI, "fairness" isn't just about ensuring every student gets a turn to use the tech; it's about making sure the AI doesn't unintentionally favor or disadvantage any group of students. Here's how to foster fairness:

- **Bias Awareness:** AI models, including ChatGPT, can sometimes hold biases because they're trained on vast amounts of data from the Internet. Recognize this fact and be prepared to address or correct any biased information or suggestions that might emerge.
- **Diverse Training Data:** If you're using AI tools that allow for customization, or if you're exploring other AI educational tools, inquire about the diversity of the training data. The more diverse the data, the more inclusive and fairer the AI tool is likely to be.
- **Accessibility:** Ensure the AI tools you use are accessible to all students, including those with disabilities. For instance, tools should have options for larger text, voice outputs, or compatibility with screen readers.
- **Cultural Sensitivity:** Always be cautious that the AI tools don't propagate stereotypes or show cultural insensitivity. This means regularly checking their outputs and discussing with students the importance of respecting all cultures.
- **Feedback Loops:** Use AI tools that allow for feedback. If a tool provides an incorrect or unfair response, there should be a way to report and correct it. Companies like

OpenAI value user feedback to improve their systems continually.

- **Inclusive Testing:** Before fully integrating an AI tool, test it with a diverse group of students. This can give insights into whether the tool has any inadvertent biases or is disadvantaging any particular group.
- **Constant Dialogue with Students:** Open a dialogue about fairness in AI. Encourage students to speak up if they feel a tool is being unfair or if they don't understand a response. Their feedback can be invaluable in ensuring an equitable learning environment.

Reflection Moment: *Think about the AI tools you've encountered or used so far. Are there moments when they seemed to favor certain students or provide responses that felt biased? How can you address these situations in the future?*

Through careful consideration and active engagement, teachers can harness the power of AI in ways that foster a fair, inclusive, and empowering learning environment for all students.

GUARDING ACADEMIC INTEGRITY IN THE AGE OF AI

In today's tech-infused world, just as educators are leveraging AI to enhance learning experiences, students might be tempted to use these tools as shortcuts, bypassing genuine learning. It's akin to the age-old tug-of-war between innovation and its misuse. With challenge, however, comes opportunity. As educators, we can employ AI to create assignments that are more analytical, open-ended, and tailored to individual students, making the traditional concept of "cheating" obsolete. Platforms like Turnitin.com now harness AI to detect plagiarized content, offering another layer of defense.

Furthermore, by integrating real-time collaborative tools and fostering classroom discussions on ethics and the importance of genuine learning, we can guide students toward a deeper understanding and appreciation of knowledge. In the AI era, the key is not to eliminate the tool but to transform the task.

Reflection Moment: *Think back to a time when you suspected a student might have taken a shortcut on an assignment. How might the situation have been different if the assignment were designed with AI's capabilities in mind? How can you transform future tasks to prioritize understanding over rote completion?*

Ruthie and I were reminiscing about our early teaching days, chuckling over the antiquated methods we once thought were groundbreaking. We debated a question that has become all the more pertinent in the age of AI: "How do we know if students truly understand the material, rather than just regurgitating pre-fabricated answers from the web?" Ruthie, in her signature wisdom, mused, "Maybe we're not challenging them in the right way. Instead of giving them time to prep and possibly seek external 'help,' why not put them on the spot?"

Inspired by this, we devised an activity that not only tested genuine understanding but also addressed the challenge of potential AI shortcuts. We called it the "Instant Expert" Presentation.

THE "INSTANT EXPERT" PRESENTATION

Objective: To assess students' genuine understanding of a topic and their ability to think on their feet, as well as their presentation skills.

Topic Selection: At the start of the lesson, provide each student with a sealed envelope. Inside each envelope is a unique topic related to the subject at hand. Topics should be diverse, ranging from easy to more challenging, but all should be within the scope of the current curriculum.

Research Time: Allocate a strict time period (e.g., 20 minutes) during which students can research their given topic. They can use any online resources, including AI tools, to gather information. The goal is not just to gather data but to understand it.

Presentation Preparation: Once research is complete, give students another fixed time limit (e.g., 15 minutes) to prepare a brief presentation on their topic. They can create quick digital slides, write bullet points, or design a poster—whatever helps them convey their understanding. No further online research is allowed in this phase.

Stand and Deliver: After preparation, students are randomly selected to present their findings to the class. Each presentation should be brief (e.g., three to five minutes), followed by a Q&A session where both the teacher and fellow students can ask questions.

Assessment: Grade students not just on the content of their presentations but also on their ability to articulate their understanding, answer questions on the spot, and effectively synthesize their research in the limited preparation time.

The "stand and deliver" approach gives teachers immediate

insight into a student's genuine understanding. There's no overnight period to allow for potential additional assistance or further online aid.

Since the task demands immediate research, synthesis, and presentation, students cannot solely rely on prefabricated online answers. They must genuinely understand the material to answer spontaneous questions.

Beyond content knowledge, this task fosters skills like quick thinking, synthesis of information, and public speaking.

This method shifts the emphasis from mere content regurgitation to authentic understanding, comprehension, and application. By requiring students to immediately stand and deliver, educators can cultivate genuine learning and deter shortcuts. The task also subtly underscores the importance of genuine understanding over simple information retrieval.

As we close the chapter on safeguarding our classrooms and ensuring fairness in AI, a foundational piece emerges on the horizon: the art of crafting effective prompts. The success of integrating AI into our classrooms hinges significantly on how we communicate with it.

Just as a seed needs the right conditions to flourish, AI needs the right prompts to deliver its best results. In Chapter 4, we will not only examine the intricacies of prompting, but we will also use our "Instant Expert" example as a hands-on guide. This will serve as a blueprint for mastering the art of the prompt. Through it, we will build your toolbox, ensuring that every interaction with AI is meaningful, precise, and fosters authentic learning. Prepare to delve into the core of AI's potential, refine your prompting skills, and set the stage for enhanced learning experiences!

CHAPTER 5

CRAFTING THE PERFECT PROMPT—RISING TO THE OCCASION

The rhythmic kneading of dough, the warm aroma filling the kitchen, and the anticipation of that first slice of freshly baked bread–it's a process that's both comforting and rewarding. On a cold, rainy weekend, I felt inspired to relive those moments from my childhood when the house would be filled with the scent of my grandmother's bread.

Eagerly, I got all my ingredients ready, and following a recipe I'd found, embarked on my bread-making adventure. While everything appeared straightforward, I soon encountered the tricky part–adding the yeast. This tiny ingredient, seemingly so minuscule in the grand scheme of the recipe, held monumental power.

My first batch, where I underestimated the yeast, gave me a dense, flat loaf. Determined, I tried again, but this time I overcompensated and ended up with an overflowing mess in my oven.

Drawing parallels between bread-making and crafting AI prompting, I realized that the same principles apply. It's all about balance, specificity, and understanding the impact of each element.

THE TENETS OF GOOD AI PROMPTING:

Precision is Key

Just as too little or too much yeast can ruin a bread batch, the specificity of your prompt can drastically affect the AI's response.

Open-ended vs. Direct Questions

Open-ended prompts like, "Tell me about Renaissance art" can yield a broad response. For our "Instant Expert" lesson, something more specific like, "What were the main artistic techniques used during the Renaissance?" can guide the AI to a more tailored answer.

Limitations and Boundaries

It's essential to set boundaries. If you're not specific, you might end up with too much information, much like the over-rising bread. For instance, "Give a five-minute presentation on Leonardo da Vinci's impact on the Renaissance" sets a clear limitation.

Active Engagement

Encourage students to ask follow-up questions. This iterative process, much like the repeated kneading of dough, can refine the output and deepen understanding.

Checking for Bias and Accuracy

Just as you'd inspect your ingredients, always review AI-generated content for any unintentional biases or errors. Guide students to do the same.

Customization for Different Learners

Some students might need more detailed prompts, while others benefit from broader ones. It's similar to adjusting a recipe based on dietary needs.

Having navigated the tenets of effective AI prompting, we're now equipped with a richer understanding of how to communicate with AI in education. It's time to put these principles to the test. Drawing on our "Instant Expert" lesson, we'll embark on a practical exploration, using each of the six tenets as our guiding stars. By melding theory with hands-on application, we'll unveil the potential of crafting impeccable prompts that cater to our educational needs. Let's see how these principles breathe life into our lesson, elevating it to new heights of interactive learning.

Using our "Instant Expert" lesson as a practical application, let's investigate each tenet. Imagine asking the AI for information on a topic like "The Solar System." Depending on how you phrase your prompt, the depth, breadth, and focus of the information can vary dramatically. This lesson will serve as our guiding example, helping us master the art of prompting, ensuring our "bread" rises perfectly each time—with students fully engaged, informed, and inspired.

NAOMI'S JOURNEY INTO THE SOLAR SYSTEM

Ruthie's granddaughter Naomi's eyes widened in

excitement as she drew the folded paper from the jar, unfurling it to reveal her assigned Instant Expert topic: the Solar System. As a 7th grader with a deep fascination for the night sky, this was the topic she had secretly hoped for. Yet, the vastness of the subject was daunting. How would she capture the grandeur and detail of our cosmic neighborhood?

Noticing her mixed emotions, Ruthie approached her with a comforting smile. "The Solar System is so vast and fascinating," Ruthie began, "but with the right questions and some AI magic, you'll not only be able to break this down to manageable parts to understand it, but you'll also dazzle the class with your presentation."

Feeling a mixture of nervousness and excitement, Naomi was ready to start her journey. With Ruthie's guidance, their collaborative exploration began.

"Let's initiate with a comprehensive overview," Ruthie suggested, illustrating how to draft a clear and succinct request.

"How about asking the AI for a brief summary of each planet in the Solar System?"

Typing carefully, Naomi entered, "Can you provide a concise summary of each planet in the Solar System?" Within moments, a neatly outlined and digestible set of descriptions was on the screen.

Ruthie gently nudged her forward, "Now, let's take it further. Pick a planet you're particularly intrigued by."

Without hesitation, Naomi replied, "Saturn! Those rings have always captured my imagination."

Ruthie smiled, emphasizing the need for specificity. "Perfect choice! Ask about Saturn's rings, their composition, and the mysteries they hold."

As Naomi continued her interaction with the AI, her initial apprehension transformed into genuine enthusiasm. Ruthie continually reminded about the importance of precision, prompting her to scrutinize the information for relevance and accuracy.

Whenever they encountered an interesting tidbit or puzzling concept, Ruthie encouraged further probing or verification against trusted sources.

With our team's mentorship and the AI's vast knowledge, Naomi and the other students in our summer trial attacked their topics with vigor. By the end of the class period, Naomi not only had a solid grasp of the Solar System, but she also developed a keen sense for effective research and inquiry.

The following day, as Naomi stood confidently in her Zoom square, her presentation was a synthesis of passion, knowledge, and clarity. Every detail of Saturn and its rings was conveyed with enthusiasm, and each query from her fellow AI explorers was met with a well-informed response. The results were a testament to the enriching work spent with WISDOMK12 mentors and their AI-powered assistant.

RUTHIE'S AI PROMPTING PROCESS

Clarity in Inquiry:

- Ruthie's Suggestion: Start with a comprehensive overview.
- Implementation: Naomi asked for a "concise summary

of each planet in the Solar System," leading to a structured, understandable response.

Specificity:

- Ruthie's Suggestion: Delve deeper into a topic of interest.
- Implementation: Naomi chose Saturn and then specifically asked about its rings, their composition, and the mysteries they hold.

Precision:

- Ruthie's Emphasis: Ensure every question is well-defined to obtain relevant information.
- Implementation: Naomi refined her queries, honing in on the exact details she wanted, like the composition of Saturn's rings.

Cross-Verification:

- Ruthie's Strategy: Scrutinize the information for relevance and accuracy.
- Implementation: Whenever they encountered a fascinating detail, Ruthie encouraged Naomi to verify it against trusted sources, ensuring its accuracy.

Iterative Exploration:

- Ruthie's Method: Encourage further probing when faced with interesting or puzzling concepts.
- Implementation: By repeatedly interacting with the AI, Naomi deepened her understanding, gradually piecing together a detailed picture of the Solar System.

Reflect and Synthesize:

- Ruthie's Approach: After gathering information, take a step back and synthesize the knowledge.
- Implementation: Naomi prepared her presentation by weaving together the various facts and details, resulting in a cohesive and captivating narrative.

Prompting Power!

Ruthie had always believed in the power of diverse learning tools, and after witnessing the success of our AI-driven "Instant Expert" lesson with Naomi and so many others in our trial, she was determined to integrate this technique more comprehensively into her ELA and literature lessons. It was brainstorming time.

Adaptable Prompting for Varied Learning Styles

Understanding the varied learning needs of her students, Ruthie initiated a project on Harper Lee's *To Kill a Mockingbird.*

For her visual learners, Ruthie prompted the AI: "Provide an infographic detailing the timeline of events in *To Kill a Mockingbird.*" The outcome was a vivid representation of the novel's sequence, helping her visual learners understand the flow and progression of the story.

Ruthie's auditory learners were assigned a different task. Using the prompt, "Summarize Chapters 1 to 3 in *To Kill a Mockingbird* in a conversational tone, highlighting key events," Ruthie received a simplified version suitable for a podcast-style delivery, making the tale more approachable for students who struggled with the dialect, allusions, and metaphors of the 1930s setting.

For the kinesthetic learners, a drama activity was

organized. With the prompt "List interactive activities related to *To Kill a Mockingbird* in which students can re-enact key scenes," Ruthie obtained a structured approach for students to physically engage and embody the characters.

Curating Content for Different Age Groups

Ruthie had been preaching about the value of ChatGPT to teachers across grade levels, subject areas, schools, and even states, so she frequently had to tailor her prompts accordingly when friends asked for ideas. For elementary students, she prompted: "Narrate a child-friendly version of the Aesop's fable 'The Tortoise and the Hare.'" The AI generated understandable tales that captivated the young minds.

In Ruthie's friend's 8th grade honors class, they were exploring narrative structures. Prompting, "Provide a step-by-step breakdown of the Hero's Journey in *The Hobbit*," Ruthie received a detailed guide that students dissected and discussed.

For high school students, deeper textual analyses were encouraged. A U.S. History teacher, Nate, included *The Great Gatsby* in his tenth grade honors class and was open to suggestions for strategies that literature teachers used to analyze the text. Ruthie asked Nate if he had looked into literary elements such as tone or imagery. Nate recalled that years earlier, a question on the state test focused on the tone of an historical document, and Nate's students had not performed well. "We discuss theme, but that's about it," admitted Nate.

Ruthie prompted ChatGPT: "Discuss the symbolism in *The Great Gatsby*." The list of symbols and their explanations emphasized the novel's exploration of the American dream, the hollowness of wealth, the moral decay of society, and the elusive nature of love and happiness. The specificity of the annotated list would enable Nate to discuss the excesses and disillusionment of the Roaring Twenties in a different way.

Multimedia Integrated Prompting

Understanding the value of multimedia, especially in the English classroom, Ruthie prompted: "Suggest five YouTube videos that provide comprehensive analyses of Jane Eyre." This list served as supplementary material, offering another teacher's students diverse perspectives on the classic novel.

Prompting for Real-World Applications

While exploring persuasive writing, Ruthie posed the question: "How are rhetorical devices used in modern advertising?" This prompt's outcome provided examples linking literature and real-world applications, making the lesson immediately relevant to her students.

Collaborative Prompting for Group Work

As part of a group project on contemporary poetry, Ruthie crafted the prompt: "Outline a collaborative analysis project on Maya Angelou's poem 'Still I Rise,' detailing roles, tasks, and presentation guidelines." This became the framework for her students' teamwork, emphasizing collective research and discussion.

Self-Assessment and Feedback Generation

Ruthie's innovative application of AI in her English classes set her apart as an educator. Her lessons were not just about reading and writing; they were about understanding, experiencing, and applying. Such lessons were time-consuming to construct. Through AI, Ruthie was now able to quickly transform her classroom into an interactive hub of tailored learning, ensuring every student felt seen, heard, and catered to.

When educators master the intricacies of forming precise and purposeful prompts for ChatGPT, they unlock a new dimension in teaching, transforming passive learning into an active exploration. Good prompting empowers students to dive deeper, encourages tailored learning experiences, and fosters an environment where every question is a gateway to a treasure trove of knowledge. It bridges the gap between conventional teaching and futuristic learning, making education a collaborative dance among teacher, student, and technology.

Ruthie's mastery of prompting didn't just enhance her teaching; it revolutionized her classroom. As I pondered my own revelations about the art of "power prompting" for ChatGPT, inspired by an afternoon of bread-making misadventures, I recognized its transformative potential for educators globally. This collective insight between Ruthie and me became our beacon: ChatGPT could redefine education, making it truly student-centric. With the vision of fostering engaging, adaptive, and differentiated learning for every student, we passionately embarked on our journey to create WISDOMK12, a tool to harness the immense power of ChatGPT and make teaching joyful again!

Here are some additional considerations when creating power prompts for ChatGPT, along with examples:

Ask Open-Ended Questions: Avoid yes-or-no questions to encourage more detailed and comprehensive answers.

- *Example:* Instead of "Was Shakespeare important?" ask "Why was Shakespeare considered a significant figure in literature?"

Specify the Format or Structure: If you have a particular format in mind for the answer, be sure to specify it.

- *Example:* "List three main causes of the French Revolution in bullet points."

Guide Depth and Detail: Indicate whether you want a brief overview or a thorough explanation.

- *Example:* "Provide a short summary of photosynthesis." vs. "Explain photosynthesis in detail, including the light-dependent and light-independent reactions."

Set Boundaries: If there's information you want to exclude or focus on, be sure to state that.

- *Example:* "Describe the plot of Moby Dick without giving away the ending."

Ask for Comparisons or Analogies: This can make complex topics more understandable.

- *Example:* "Explain the concept of relativity using a train analogy."

Encourage Creativity: If you're looking for a more creative or out-of-the-box answer, indicate that.

- *Example:* "Imagine if trees could talk. Write a dialogue between an oak tree and a pine tree discussing the seasons."

Clarify the Intended Audience: This ensures the answer is tailored appropriately in terms of complexity and tone.

- *Example:* "Explain the water cycle as if you were teaching it to a group of third-grade students."

Indicate the Desired Tone: Whether you want a formal explanation, a humorous take, or a casual discussion can influence the response.

Example: "In a humorous tone, describe the challenges of being a medieval knight."

Request Examples or Illustrations: When you want the model to provide practical applications or illustrative scenarios.

- *Example:* "Explain the principle of supply and demand and give a real-world example of its impact."

Seek Multiple Perspectives: Encouraging the model to provide various viewpoints or arguments can give a balanced overview.

- *Example:* "Outline the pros and cons of using nuclear energy as a primary power source."

Mixing and matching based on your needs can guide ChatGPT to generate the most useful responses. And as always, experimentation is key; playing with different phrasings can yield a range of outputs, giving you a sense of how best to interact with the model for your needs.

CHAPTER 6

EMBRACING THE WAVE—RIDING AI TO PERSONALIZE LEARNING

The public narrative around AI in education has often teetered towards fear. Headlines scream about the dangers of machines replacing the human touch, painting a bleak picture of cold, mechanized classrooms where the sacred bond between student and teacher would become obsolete. But Ruthie and I stood perplexed amidst this tumultuous backdrop. In our experiences, we didn't see the need for such fears.

In the months since we began our quest to best reimagine our educational practices in the Age of AI, we found very few teachers who had given serious thought to the enormous promise of AI in education. Some were willing to jump into the conversation with opinions formed before ever trying it. Here are a few excerpts from the wide-ranging conversations we had with fellow teachers:

- "We need a policy ASAP to prevent students from using ChatGPT at home."
- "I read plenty about it, but it can't really help me teach French."
- "I told my students that software is rapidly being developed to detect the use of ChatGPT."
- "I know all about it. I had a conversation this summer with my cousin's husband who works with computers."
- "I haven't used it yet, but I volunteered to be on a committee to write a policy about use of ChatGPT and other artificial intelligence. We'll soon see."
- "I quit giving homework that requires written answers because there's no way to tell if a clever student used it and then added a few touches to make it look like it was their work."

An article in *Education Week* by Arianna Prothero (September 29, 2023) caught our interest: "How Students Use AI vs. How Teachers Think They Use It, in Charts." In a study conducted by the Center for Democracy & Technology, June–August 2023, a survey of 1,005 middle and high school teachers revealed that 90% of teachers surveyed said that they thought their students had or had probably used generative AI for school. In reality, although 58% of students reported that they had used generative AI, only about a quarter of students said they had used a program like ChatGPT for school assignments. Students were more likely to have used it for personal use, searching topics ranging from hobbies to health and anxiety issues. An unexpected finding, however, was that although only 58% of the general student population reported that they had used AI, 72%

of students with an IEP or 504 plan had used generative AI. Though requiring further study to validate, we hypothesize that the collaborative way in which special education teachers and their students work to find alternative avenues to meet specific needs emboldens students with IEPs or 504 plans to embrace new learning opportunities."

Your personality type has much to do with the way you are likely processing what you have heard about AI in education. (If you've never thought deeply about your own personality or haven't had the opportunity to take the Meyers-Briggs Type Indicator or explore the Enneagram of Personality, take the free test at 16personalities.com. It's enlightening and quite valuable in terms of understanding others.) Ruthie and I have similar personalities, which is what drives both of us to stay on the leading edge of our profession. We are classroom teachers, but we are also learning research scientists. We're not "Waiting on the World to Change," as John Mayer sang, but instead are eager to investigate ways to strengthen our craft. We care deeply about our students and our schools and seek the best possible resources. Teachers are learners, but we are also sharers, which is why we were motivated to write this book.

LEARNER + SHARER = TEACHER

We have concluded that "The Era of Technology" or "The Digital Age" are not sufficient labels for current times. "The Age of Artificial Intelligence" is where we are now, ready or not. Sticking our heads in the sand and teaching the way we have been teaching, even if we are integrating standard technology in the classroom, will no longer suffice.

In an article published in *Scientific American* (February 10, 2023), John Villasenor asserts:

> To remain competitive throughout their careers, students need to learn how to prompt an AI writing tool to elicit worthwhile output and know how to evaluate its quality, accuracy and originality. They need to learn to compose well-organized, coherent essays involving a mix of AI-generated text and traditional writing. As professionals working into the 2060s and beyond, they will need to learn how to engage productively with AI systems, using them to both complement and enhance human creativity with the extraordinary power promised by mid-21st- century AI.

AI has been and will continue to be misused by students and people in all types of careers who simply want to let it do the work *for* them; we are embracing it as an incredibly powerful assistant who will do the work *with* us and with our students.

We had witnessed first-hand the magic that well-crafted prompts could conjure in the realm of teaching. We had seen the power of AI in sparking student curiosity, in making lessons more relevant, and in pushing the boundaries of what was possible in a classroom. This wasn't just about technology; it was about reimagining the very nature of teaching and learning.

Our belief was counterintuitive to the prevailing winds of fear: AI, when used right, could amplify the human connection in classrooms rather than diminish it. Instead of replacing educators, it could empower them. Instead of alienating students, it could draw them closer, catering to their unique needs and interests.

Ruthie often mentioned an instance where, after introducing the art of precise prompting with ChatGPT, she overheard two students animatedly discussing the French Revolution. She thought she might be hallucinating, but it was actually happening. And their debate wasn't just a mere repetition of facts, but instead was infused with insight, curiosity, and a thirst for deeper understanding. They weren't retreating from personal interaction; they were more engaged than ever.

The core of our philosophy is simple: The wave of AI in education is inevitable, but like any tool, its impact depends on how it is wielded. If educators approach AI with fear, viewing it as a competitor that could replace them or a cheat tool that will prevent their students from learning to write, its true potential will remain untapped. If they see it as a collaborator, though, they can harness its strengths to address the very real challenges that have long plagued education: student disengagement, the one-size-fits-all model of teaching, and the inability to provide personalized attention.

In our conversations, we often circle back to a fundamental question: What if, instead of being the end of meaningful interactions in the classroom, AI is the catalyst that brings about more profound, more enriched dialogue between students and teachers?

The possibilities are tantalizing. Imagine classrooms where teachers, armed with insights from AI, cater to the individual learning needs of every student. Imagine students being able to probe deeply into topics of their interest, driven by their own inquiries and aided by the vast knowledge reservoirs of AI.

The wave is upon us, and the choices are clear: We can either stand our ground and get swept away, or we can ride it, steering it purposefully towards a brighter, more inclusive future of education.

Ruthie and I chose the latter. We believe that with the right mindset and tools, e.g, WISDOMK12, we will guide educators to harness AI's transformative power, turning classrooms into vibrant hubs of exploration, dialogue, and discovery.

As Ruthie and I investigated the capabilities of ChatGPT, we stumbled upon a profound revelation. While AI, especially ChatGPT, was an excellent tool for responding to students, it was also an astute observer. Every interaction, every response a student gave, was an opportunity for ChatGPT to learn, adapt, and fine-tune its subsequent interactions. This real-time feedback loop provided a rich tapestry of information about each student.

STUDENT WRITING TASKS AND CHATGPT: WHAT CAN BE UNEARTHED?

1. **Gauging Writing Proficiency:** As students responded to writing prompts, the structure, vocabulary, and fluency of their responses offered insights into their writing capabilities. This wasn't just about catching grammatical errors or checking for coherent sentences. It was a deeper dive into understanding whether the student was able to articulate thoughts cohesively, use varied sentence structures, and employ vocabulary appropriately.
2. **Deciphering Reading Levels:** From a student's interactions, it was possible to infer their reading level. If a

student consistently misunderstood complex passages or struggled with higher-level vocabulary, it was a sign they might benefit from reading materials tailored to their level. Conversely, if they easily navigated through dense texts, it was an indicator that they were ready for more advanced readings.

3. **Pinpointing Interests and Strengths:** The topics that a student chose to explore, the questions they asked, and the manner in which they engaged with the content provided glimpses into their interests and strengths. A student showing immense curiosity about space exploration might have a passion for astronomy, and this could be harnessed in future lessons.
4. **Identifying Areas of Improvement:** ChatGPT's interactions could highlight areas where a student might need more support. This could range from conceptual misunderstandings to challenges in articulating a particular argument or response.
5. **Feedback and Iteration:** One of the strengths of ChatGPT was its ability to provide instant feedback. This immediate loop allowed students to rework their writing in real-time, incorporating suggestions and refining their skills.
6. **Personalizing Future Interactions:** By learning from previous interactions, ChatGPT could personalize future prompts and resources. If a student showed a keen interest in historical events, future prompts might revolve around historical contexts, ensuring the student

remains engaged and challenged.

Ruthie began integrating these insights into her teaching methodology. Instead of one-size-fits-all assignments, she started tailoring writing tasks based on what she understood about each student from their interactions with ChatGPT. For some, this meant delving deeper into areas of passion. For others, it meant providing additional resources or scaffolds in areas in which they struggled.

Moreover, this wasn't a static process. As students evolved, so did their interactions with ChatGPT, ensuring that their learning experiences remained dynamic, relevant, and personalized.

The potential was revolutionary. Instead of educators spending countless hours trying to decipher each student's needs, strengths, and challenges, they now had a tool that could assist in this process, making the path to personalized education smoother and more efficient.

To Ruthie and me, this was more than just a technological advancement; it was a paradigm shift in teaching. By leveraging the power of AI, educators could now truly meet students where they were, ensuring that learning was not just personalized but also deeply meaningful.

STRATEGIC STUDENT ENGAGEMENT

As educators embark on this journey of leveraging AI to foster student growth, it's imperative to kickstart the process with strategic engagement. The initial interactions a student has with ChatGPT are pivotal, as they lay the foundation for the AI to understand the student's strengths, challenges, and interests. For teachers eager to jump in, here's a list of six writing prompts that can help facilitate these foundational interactions. These are written specifically for 8th graders. Remember, you can get specific and ask ChatGPT to craft writing tasks for whatever level you are teaching.

1. **Historical Perspective:** "Imagine you are living during the American Revolutionary War. Write a diary entry detailing a day in your life, describing the challenges you face and your feelings about the ongoing war."
2. **Science Exploration:** "If you were given a chance to travel to any planet in our solar system, which one would you choose and why? What do you hope to discover or experience there?"
3. **Personal Reflection:** "Think about a time when you faced a significant challenge or obstacle. Describe the situation and how you overcame it. What did you learn about yourself in the process?"
4. **Creative Endeavor:** "Write a short story starting with the sentence: 'As the clock struck midnight, the entire town gathered around the ancient oak tree, waiting for the legend to come to life.'"
5. **Societal Inquiry:** "In your opinion, what is the most pressing issue facing our society today? Explain why you believe it's critical, and propose a solution or steps that could address the problem."
6. **Literary Analysis:** "Choose your favorite character from a book you've recently read. Describe their personality traits and motivations, and discuss why you find them intriguing."

By prompting students with these diverse topics, ChatGPT will gain a rounded view of the student's writing capabilities, areas of interest, and personal experiences. This foundational understanding will set the stage for even more tailored and effective interactions in the future.

GETTING THE MAX OUT OF MAX

Max, a quiet 8th grader, often kept to himself in Ruthie's English class. Though he performed reasonably well in most tasks, Ruthie felt she couldn't tap into his true potential. The introduction of ChatGPT presented an excellent opportunity to rootle into understanding Max.

On the **first prompt** about the American Revolutionary War, Max's response was vivid, detailing a day in the life of a young blacksmith's apprentice named Eli. From his piece, it became evident that Max had a deep appreciation for historical contexts, particularly those revolving around craftsmanship and trade.

For the **second prompt**, Max's choice was Saturn. He expressed a wish to glide among its rings and discover if any hidden moons existed. Through this, it was clear he harbored an affinity for space exploration, combined with a poetic imagination that Ruthie hadn't noticed before.

Max's **personal reflection** was an account of his recent relocation to this school. He wrote about the challenge of leaving old friends behind and the struggle to fit into a new environment. This was an eye-opener for Ruthie. Max's reticence now had context.

Max's response to the **creative endeavor** revolved around

a town where every midnight, residents would receive wisdom from the ancient oak tree. The tale was rich in metaphors, symbolizing guidance and seeking answers from the past.

For the **societal inquiry**, Max chose environmental degradation. His solution revolved around community gardening and reforestation projects. This showcased his keen interest in the environment and the practical solutions he envisaged for real-world problems.

Lastly, in the **literary analysis**, Max chose Samwise Gamgee from *The Lord of the Rings*. He admired Sam's loyalty, bravery, and underlying wisdom. Through this, Ruthie could decipher Max's value for steadfastness and unwavering support in friendships.

From ChatGPT's analysis, several insights were shared with Ruthie:

1. **Historical Affinity:** Max's love for history, particularly around crafts and trades, could be tapped into by recommending books or documentaries on ancient civilizations and their craftsmanship.
2. **Space Fascination:** This could be an avenue to introduce Max to speculative fiction, allowing him to explore and expand his poetic imagination further.
3. **Personal Challenges:** Knowing Max's recent relocation struggle, Ruthie could facilitate buddy systems or group projects, helping him forge new bonds in class.
4. **Environmental Concern:** Ruthie might guide Max towards the school's environment club or encourage him to initiate community projects.
5. **Value Systems:** Understanding Max's appreciation for

loyalty and bravery can help Ruthie design character analysis tasks, debates, or discussions around these themes.

Ruthie, equipped with these insights, felt better prepared to cater to Max's needs, interests, and strengths. With the AI's assistance, she was ready to design lessons that would not only engage Max but also empower him to flourish in his academic pursuits.

After Max submitted his responses to the writing prompts, ChatGPT didn't just stop at understanding his interests and personal experiences. The system has been designed to comprehend the structural intricacies, vocabulary usage, and the depth of thought presented in the content, thus gauging a student's reading and writing levels.

The process of deducing reading levels using AI revolves around a combination of linguistics, data analysis, and machine learning. Here's a glimpse into the science behind it:

1. **Lexical Analysis:** ChatGPT first breaks down the text into its constituent elements, examining word choices and vocabulary. Using extensive databases, it matches the vocabulary with grade-appropriate word lists. For instance, the use of more complex, multi-syllabic words or specialized terms may indicate a higher reading level.
2. **Sentence Structure and Complexity:** The system evaluates sentence length, variety, and complexity. Longer, compound-complex sentences can hint at a more advanced writing capability. Conversely, shorter, simpler sentences might indicate a more basic reading/writing level.

3. **Thematic Analysis:** By understanding the themes and depth of content in the writing, the AI can gauge the maturity of the student's thoughts. A nuanced discussion of intricate topics can signify advanced comprehension
4. **Grammar and Conventions:** Proper usage of grammatical structures, punctuation, and varied sentence starters can give insights into a student's grasp of the language and their formal writing skills.
5. **Comparative Analysis:** *The AI has access to vast* datasets of writings across different age groups and grade levels. By comparing a student's submission with this dataset, it can approximate where the student's abilities lie on a spectrum of reading and writing proficiencies.
6. **Machine Learning Feedback Loop:** The more data the AI processes, the better it gets at making these determinations. Over time, as AI interacts with more student responses, its predictions become increasingly accurate, refining its understanding of reading levels.

By synthesizing the insights from these analyses, ChatGPT can provide a fairly precise estimation of a student's reading and writing caliber, allowing educators to adapt their instruction accordingly.

From Max's detailed response about Eli, the young blacksmith's apprentice, ChatGPT picked up on his use of specific historical terminologies, intricate sentence structures, and mature vocabulary. The narrative showed a reading comprehension and writing ability that was slightly above the average 8th-grade level.

Max's imaginative thoughts about Saturn and his eloquent

tale about the wisdom-sharing oak tree further underscored his linguistic strengths, reflecting a flair for both non-fictional knowledge and creative storytelling.

On the other hand, Max's personal reflection about relocating revealed a simpler, more direct style, potentially hinting at areas where he could work on personal narrative and emotional expression in writing.

Ruthie realized that by analyzing responses from her students, ChatGPT could provide her with a spectrum of reading and writing proficiencies. Some students, like Max, showcased advanced capabilities, while others were at par or slightly below the expected grade level. This information was invaluable to Ruthie. It meant she could truly personalize her teaching strategy.

Equipped with this data, Ruthie is looking forward to creating a reading list for each student. For Max, she picked titles that resonated with his historical interest, like *Crispin: The Cross of Lead* by Avi, and books that indulged his cosmic curiosity, such as *A Wrinkle in Time* by Madeleine L'Engle. Knowing his current reading level ensured that while these books would challenge him but not overwhelm him.

This methodology not only caters to each student's unique needs but also fosters an inclusive classroom environment. Every student, regardless of their reading level, has access to literature that resonates with their interests and is comprehensible to them. By ensuring students aren't frustrated with inaccessible content or bored with too-basic material, Ruthie can increase engagement, comprehension, and the joy of reading in her classroom.

TAKING A SWIPE AT PERSONALIZED LEARNING

Having plunged into the potential of ChatGPT and witnessed its impressive ability to discern students' writing and reading levels, Ruthie and I recognized a golden opportunity. We began to imagine a world where every piece of student writing, be it an essay, a journal entry, or a quick response, becomes a valuable data point to tailor and enhance instruction. This led to the creation of our very own acronym. Everybody loves acronyms in K12, right?

STUDENT WRITING & INTEREST PROFILE ENTRIES (SWIPE)

SWIPEs aren't just a series of prompts. They represent a dynamic system where each prompt plays a pivotal role in gauging different writing skills. These skills range from narrative and persuasive writing to descriptive, reflective, and argumentative writing. As students responded to these prompts, ChatGPT meticulously analyzed the submissions. The goal was straightforward but revolutionary: understand each student's unique strengths as well as areas that required further refinement.

More than just a writing assessment tool, SWIPEs become the conduit for individualized learning. By offering insights into students' proclivities, preferences, and proficiencies, teachers can generate subsequent prompts through ChatGPT that resonate with individual learners, pushing them gently towards growth while ensuring they remain engaged and inspired.

In our vision, SWIPEs aren't merely tasks or assignments. They embody the symbiotic relationship we envision between AI and the classroom. For true learning personalization to materialize, it is imperative for the tool (ChatGPT), the teacher, and the student to operate in harmony, each learning from and adapting to the other. With SWIPEs, we believe we have set the stage for precisely this kind of transformative, individualized learning experience.

At its core, the idea behind SWIPE was analogous to the simple act of swiping a card at a store or an ATM. When you swipe a card, it's not just a physical action; it signifies the transfer of information. The magnetic strip on the card carries critical data about the account holder, and in a matter of seconds, this data is read, processed, and appropriate actions are taken based on it. Similarly, with every SWIPE entry a student completes, they are transferring a wealth of information about themselves—their writing style, comprehension, thought process, and even their interests and aspirations.

This constant flow of data is inestimable. Just as a card swipe ensures a seamless transaction by communicating essential information between the card and the machine, SWIPE facilitates a seamless exchange of insights between the student, the teacher, and the AI. The AI learns from the student's writing, the teacher gains insights from the AI's analysis, and the student benefits from the personalized feedback and resources based on this combined understanding. However, having a system that gathers information is only half of the equation.

While SWIPEs are only one part of our revolutionary AI-assisted approach to learning, they are a KEY reason that learning with WisdomK12 is an evolutionary process . . . the supports evolve in parallel with student growth.

The real magic lies in how this information is utilized. How can a teacher harness these insights effectively? How can the student act on the feedback to foster growth? And how does the AI evolve its approach based on the continuous stream of SWIPEs? These questions set the stage for our exploration into the next crucial component of our system: the student/teacher feedback loop.

In a later chapter, we'll explore this feedback mechanism, understanding its significance and the transformative potential it holds for modern classrooms. It's a dance of continuous learning, adaptation, and improvement, and the rhythm is set by the knowledge transferred with each SWIPE.

CHAPTER 7

ONE SCHOOL'S BEGINNING

A small group of administrators and teachers at Ruthie's school met for a planning meeting before opening day. After the usual chatting about how the summer flew by, the principal asked, "Are you ready?"

Ruthie responded sincerely with a knowing smile, "I am more ready and excited to start school than I have ever been."

"With all your traveling, that's not what I expected to hear, but great! Why this year?"

"ChatGPT changed my teaching life, and I can't wait to share!" Then she pointed to her feet and added, "Oofos Recovery Sandals helped as well."

Ruthie shared the short version of her work with the group, and then they turned to the agenda. One of the items was looking ahead to Catholic Schools Week in January. Ruthie opened her laptop, logged into ChatGPT, and said, "Give me one minute first." As quickly as she could type "Generate ideas for celebrating Catholic Schools Week at a K-8 school," dozens of ideas began to appear. No one in the group had seen ChatGPT in action, so they were all oohing and aahing.

Later in the meeting, Principal Jenny said she'd like to update the school's alumni newsletter. Ruthie typed "Ideas for a private elementary school alumni newsletter" into ChatGPT, and a list of twenty-five possible features appeared, a few of which they marked for further exploration.

When school began, Ruthie's eighth grade ELA students were eager to talk about ChatGPT, surprised that they were "allowed" to discuss it, let alone were about to use it. "We're not going to wait around for others to tell us how and when to use it," Ruthie told her class. "You will learn *ethical* uses of artificial intelligence related to our work." Ruthie sensed that her teacher rating just went up several points.

On the first Staff Development Day, Ruthie asked for a few minutes on the junior high team agenda. Lack of time had been cited several times already that morning, as ideas were many—from monitoring hallways to hosting a nationality festival—but the veteran teachers knew that they had to budget their time.

By the time it was Ruthie's turn to talk, the meeting had already run past the scheduled time. "Lack of time has been a common theme today, so I'm happy to introduce you to something that's going to help us save time."

Knowing her colleagues were eager to get to lunch, Ruthie applied her "Show, don't tell" writing advice to the moment. That morning she had asked ChatGPT to "Create instructions for a middle school presentation on a musical genre." In the time it took to type that sentence, the app began to generate a thorough, well-organized set of instructions for this project. Ruthie tweaked about ten percent of the document to suit her personal preferences (e.g., index cards for notes) and printed enough to share with her fellow teachers. "This is a lesson that ChatGPT created in about one minute. It would have taken me much longer to start this from scratch and probably would not have been as thorough. Here is the plan that ChatGPT generated:

Exploring Music Genres

Objective: I can prepare and deliver an engaging 2–5-minute presentation on a music genre.

Materials Needed: Computer, research materials (websites, books, articles)

Steps to Create a Presentation

A presentation assignment is an opportunity for you to delve deeply into a topic as well as an opportunity to teach something to your audience. You are being given a segment of instructional time that must be taken seriously, as your presentations will provide the instruction for the unit on musical genres.

Research:

Gather information about your chosen genre from multiple reliable sources. Look for key historical events, influences, characteristics of the genre, and any interesting anecdotes.

Outline Your Presentation:

Outlining for a presentation is like pre-writing for writing tasks. Create a rough outline for yourself, including an introduction, main points, and a conclusion. Although the outline is for your own benefit, you must turn in a copy after your presentation, just to ensure that you don't skip this important step. You can use the following structure as a general guideline:

- Introduction
- History and Origins of the Genre
- Key Artists and/or Composers
- Characteristics and Instrumentation
- Notable Songs
- Impact on Society/Culture
- Conclusion (What makes this genre so special?)

Visual Aids:

Collect relevant images, videos, and audio clips to enhance your presentation. You may also bring in tangible items to support your presentation, such as record albums or band memorabilia.

Design Your Presentation:

Create visually appealing slides (IF you are creating a digital presentation) that support your speaking points. Use clear fonts and visuals, and keep wording concise. Text should be large enough for viewers to see from their seats and should not be "hidden" by busy background images.

If you are not creating a slide presentation, be sure to find or create visuals, props, and audio clips that your teacher can set up on Spotify. For example, you might create a handout or make a poster.

Prepare Speaking Notes:

You may use one or two index cards to guide you during your presentation. You should not be reading the entire presentation directly from slides, handouts, or your own notes. Presenters use visuals as props to support what they are saying.

Engage Your Audience:

Consider using interactive elements such as questions or discussions to keep your audience engaged. Know your genre well, and be prepare to field questions from your classmates.

Practice Your Delivery:

Rehearse your presentation multiple times to become comfortable with the content. Make sure you stay within the 2-5-minute time frame. Conduct a practice for a friend or a family member and solicit their feedback for improvement.

Finalize Your Presentation:

Make any necessary adjustments based on feedback from practice sessions.

Reflect on Your Presentation:

After your presentation, reflect on what went well and what you could have improved upon. You will give many presentations throughout your schooling, and for some, presentations will be an important part of their work life. Learn to design and deliver an effective presentation.

"Think about at least one thing ChatGPT could do for you. You are welcome to come to my room and use the app yourself, or just let me know what you'd like to explore, and I'll type it in. Take a little time to think about and get back to me."

Requests began to flow in by text, email, or in person.

Science teacher Tracy teaches three levels of science. "Some ideas for me would be genetic engineering, its pros and cons, and human cloning. I'd also be interested in the effects of global warming on natural & human populations & synthetic elements, that is, what they are, how they are created, and what uses they have."

Ruthie entered three separate requests for Tracy's topics. ChatGPT first generated a description of genetic engineering, then itemized its major benefits and cautions, and ended with a reminder that would be important to impart to students: "Balancing the potential benefits with the cautions is a complex task that requires ongoing research, oversight, and thoughtful decision-making,"

Next, ChatGPT introduced human cloning as "a complex and controversial topic that raises ethical, moral, and practical questions." It listed pros and cons that included observable uses (e.g., the creation of medically necessary tissues, organs, or cells); ethical concerns such as the sanctity of life; and yet unanswered questions about issues such as potential commercialization and regulatory challenges.

Finally, ChatGPT defined global warming and listed separately the effects on natural ecosystems, human populations, and synthetic elements, directly and indirectly.

The next questions came via email from Allyson, a Spanish teacher who thoughtfully immerses her students in all aspects of Hispanic culture. She also keep her students in the loop on current events related to language on a local, national, or global scale. Allyson's Day of the Dead (Dio de los Martes) celebration is always one of the highlights of the school year.

Allyson's first question for ChatGPT was, "What are the tried and true benefits of having children learn a foreign language at an early age as opposed to when they are in middle school?"

With ChatGPT sounding more like a professional conversation than a source of information, it instantly produced a list of about a dozen benefits of learning a second language at an ey arly age. In addition to benefits most people could probably brainstorm, such as increased career opportunities and a greater appreciation for cultural awareness, ChatGPT mentioned abilities discovered through research, such as improved cognitive skills and an ability to develop native-like pronunciation and fluency when exposed to a second language earlier. It also discussed other factors on which the effectiveness of early language learning depend, e.g., quality of instruction, and reminded that starting early is not the only way to become proficient in a foreign language. Allyson followed through on ChatGPT's reply, asking "If there are measurable benefits to learning a second language at an early age, why aren't we doing that across America today?" ChatGPT gave a list of key reasons within the various factors that affect instructional decisions, ranging from budget constraints to lack of consensus on which language(s) to teach to political and ideological differences in education policy across the country.

Still curious, Allyson took it further. "How many years of studying a foreign language does it take for a student to become fluent in the target language?" ChatGPT presented some general guidelines that were more about varying factors such as similarity to native language, aptitude, and teaching methods, but it didn't produce actual time estimates. Allyson revised her question by adding "On average' at the beginning, and

ChatGPT took a different approach. It explained the commonly cited framework known as the Common European Framework of Reference for Languages (CEFR) and provided a rough estimate of the time it takes to develop language proficiency at various levels, stating approximations in months to years. As we had found with other questions, ChatGPT discussed more than we had asked, again pointing out factors that cause individual progress to vary widely.

The math teachers were attending other meetings on Staff Development Day, so Ruthie thought of something cool she had seen at another school but didn't know how to implement—a Probability Fair. She remembered that younger students were given tokens and won prizes, but she didn't remember much else.

Not being a math teacher, Ruthie wasn't confident about the wording of her ChatGPT prompts for this idea, but she knew that she could tweak her question if she didn't get the hoped-for results. She first asked, "What steps would middle school students follow to set up a Probability Fair for elementary students? ChatGPT generated a list of fourteen steps with how-to details for each, from defining the objectives of the event to developing scorecards to following up with assessments to gauge what the elementary students had learned from the Probability Fair. What Ruthie still needed to know, however, was what kinds of games could be created.

Ruthie's next prompt for ChatGPT was, "What kinds of games could middle school students set up for a Probability Fair? How would they calculate the possibilities?" (It's fine to enter multiple-part questions. "Talking" to ChatGPT is just like having a conversation with a colleague.) A list of doable games appeared, each with a paragraph elaborating on how the game would look and work and the way in which probability would be calculated. A teacher could easily set up and facilitate the fair from ChatGPT's suggestions. Games involving spinners, dice, marbles, cards, board games, and jelly bean jars not only teach about probability but also make learning interactive, fun, and memorable, greatly enhancing the probability of retention.

Ruthie's assistant principal, Michelle Norris, was curious about what ChatGPT would suggest for a separate handbook for Pre-K 3- and 4-year-olds and kindergarteners. The general rules of the school's Student Handbook and Code of Conduct were not always fitting for the little ones. Michelle had already been working on a handbook but compared the results to her own work. "Wowsers!" exclaimed Michelle, as ChatGPT first produced an outline, followed by seven pages of content for a complete handbook. She found the pages useful in editing and revising her initial work.

James, our veteran social studies teacher, had been teaching enough years to have refined his curriculum to his satisfaction. He said he would be interested, though, in seeing what AI would suggest for a group project that would help middle school students understand what everyday life was like for the average American during the Great Depression. ChatGPT generated "Life in the Great Depression: A Time Capsule Project." Students would be divided into small groups, each researching a different aspect of life during the Great Depression, with topics like family life, housing, food, clothing, entertainment, work, and government programs like the New Deal. Each group would create a "time capsule" with items, images, and descriptions that reflected the aspects they research. Followed by group presentations and class discussion, the project would allow students to engage with history in a hands-on way, fostering a deeper understanding of the Great Depression's impact on the daily lives of ordinary Americans. This type of activity combines teamwork, research, and presentation skills while making history more tangible and relevant.

Holly, the school's literacy specialist, often chatted with Ruthie about literacy-related issues. Two nerds. When shown ChatGPT, Holly remarked, "This is amazing. How can I do this with a reading group or two? Maybe one with lower level kids and one with higher level kids?"

"We can brainstorm, or we can just ask the AI how it can best help us." Ruthie typed "How can ChatGPT help a reading specialist work with small groups?" The AI's response was worth repeating here, as all teachers are literacy teachers to some extent.

- **Customized Lesson Plans:** ChatGPT can generate customized plans and reading activities tailored to the specific needs and reading levels of each student in the group. By providing information about the students, such as their current reading levels and any specific challenges they may have, the specialist can receive tailored lesson plans.
- **Reading Materials.** ChatGPT can recommend or provide reading materials, including books, articles, and online resources that align with the students' reading levels and interests. It can suggest a variety of texts to keep the group engaged.

- **Lesson Content.** The specialist can use ChatGPT to assist in developing engaging and informative lesson content, including discussion questions, vocabulary exercises, and comprehension activities. ChatGPT can provide ideas for making lessons more interactive and fun.
- **Assessment Tools.** ChatGPT can help in designing and suggesting assessment tools, such as quizzes and reading comprehension exercises, to measure the progress of each student. It can also help in analyzing the assessment results and providing insights into areas that may need improvement.
- **Feedback and Evaluation:** ChatGPT can assist in giving feedback to the specialist about the group's progress and suggest strategies for improvement. It can provide recommendations based on the data and observations the specialist provides.
- **Individualized Support.** For students who require additional support, ChatGPT can generate individualized learning plans, including targeted interventions and strategies to address specific reading difficulties.
- **Professional Development.** ChatGPT can provide information on the latest research and best practices in reading instruction, helping the specialist stay updated with the field's current trends and methodologies practice reading outside of the group sessions.
- **Parent and Teacher Communication.** ChatGPT can assist in drafting communication to parents or other teachers, explaining the progress of students, areas of

concern, and recommendations for additional support.

- **Resource Recommendations.** ChatGPT can recommend educational apps, websites, and other resources that can complement the specialist's teaching efforts and help students practice reading outside of group sessions.
- **Answering Student Questions.** During group sessions, ChatGPT can assist in answering students' questions and providing additional context or explanations on various reading topics.

Holly and Ruthie continue to have AI-centered discussions. They talked about how ChatGPT can differentiate texts for students at different reading levels, like the popular NewsELA app. In this way, any teacher can use the same basic article or text and ask the AI to rewrite it at a specific reading or grade level. Students in the same classroom can be grouped according to ability level and not even realize their texts differ, as the basic information and gist of the text remains the same. Although there are few English language learners in their school, the teachers marveled at how helpful it would be for an ESL teacher to use standard upper-level texts but ask ChatGPT to adapt the wording to make it understandable to intermediate-level English language learners.

While they were chatting, Ruthie typed in, "Paraphrase the Declaration of Independence in language that a third grader would understand." And there it was, perfect for a struggling reader or a student whose second or third language is English.

A retired teacher and excellent substitute, Rod, contributed. "I see and hear things about ChatGPT replacing teachers someday. Do you think artificial intelligence will get to that point?"

"I've been working with the WISDOMK12 team all summer, and we don't think that will happen in the foreseeable future, but let's see what the AI says about itself." Ruthie asked ChatGPT, and it eloquently presented ten different areas in which AI cannot begin to match a human teacher. In general, its arguments in favor of real teachers discussed the multifaceted roles of teachers in fostering social and emotional learning, their ability to make complex decisions on the spot, and perhaps most importantly, the fact that artificial intelligence cannot replicate the human connection between teachers and students, a critical factor in the learning process and personal development.

The physical education teacher, Bob, was not at the academic team meeting where Ruthie first began the AI discussion, so Ruthie asked her husband Jay, a former phys. ed. teacher, what might be useful. He suggested I ask for a list of activities for days when students cannot change into their gym clothes, such as school picture days. ChatGPT produced a varied list of activities in which students could experience movement, ideas varying from a square dance class to relay races, chair workouts, brisk walks, and more.

Kaitlyn, the other ELA/literature teacher in the middle school, was excited about the possibilities. She asked Ruthie, "Can you give me more information on ChatGPT's capabilities? I would love to teach the students MLA format and citing research sooner than seventh grade."

In about a minute, ChatGPT had written "Citing Research in MLA Format: A Simple Guide for Sixth Graders." Kaitlyn was pleased with the document—clear, understandable, available immediately, and free of charge.

Kaitlyn continued, "I would love help with editing students' work. Is that a possibility? I would love to give them feedback sooner, but reading all their papers is time-consuming. Is that something that ChatGPT can do?"

"Yes, yes, yes!" exclaimed Ruthie. "That's what I love the most about ChatGPT! That's the work we've been doing this summer, developing an AI tool that can give students initial feedback on their work based on the instructions, rubric, and writing standards. Might you be interested in piloting the tool with me this year?"

"Sure!" replied Kate. "When can we start?"

CHAPTER 8

COUNTING ALL THE POSSIBILITIES

Ruthie was excited to be able to show Katelyn what the AI tool the team at WISDOMK12 had developed can do. She explained that they had recruited middle and high school students to write over the summer so the team could tweak the type of feedback a developing writer would receive.

"Katelyn, throughout my career, the biggest obstacle to teaching students to write has been the amount of time it takes to mark that first batch of papers. In public school, I averaged about 160 students in six classes. I tried *everything* to be able to proceed with the writing process in a timely manner."

Ruthie had gone to great lengths to be able to teach writing authentically, that is, using the necessary *process* that real writers use. In addition to forfeiting sleep, she tried alternating due dates among classes so she could at least get through one class per night. Lesson plans became quite confusing, however, especially when regular schedules were interrupted. She recruited and trained the most competent students to act as peer editors, offering community service hours in exchange for the students' extra effort, as they gave up time during their mornings, lunchtimes, or evenings editing online. When Ruthie's own children were young, she would take them along with their friends to play at school in the evenings. The novelty of playing school in a real classroom kept the kids contented while Ruthie entered grades on the protected desktop computer or switched her bulletin boards to display the most current exemplary student writing. Ruthie asked to have her planning time coincide with most students' lunch periods so they could stop in for feedback. She gave out her phone number to all students with the protocol, "Ask three before you ask me." When an occasional phone call came from a student, Ruthie would always ask which three students the caller queried first. If four students did not know something related to class, it was likely that something needed clarification. She worked hard to foster a "community of learners" where students were healthily interdependent and viewed themselves as apprentices learning a valuable craft.

For most real writers, revision is as important as the writing itself. If too much time passes between drafting and revision, however, students lose momentum and forget the point of the lessons surrounding a particular writing. Ruthie took her life's work as seriously as any professional, but the loss of personal time was a high price to pay for student success.

Ruthie showed Katelyn an essay written by Tyler, one of our team member's sons. The prompt for the writing was "Tell a personal story of failure and how you overcame it. Consider a time when you failed to accomplish something you wanted and how you responded to that failure." Tyler wrote a personal narrative about his freshman year basketball tryout.

This was WISDOMK12's response:

Analytic Understanding and Task Addressing

(Worth 28 points)

The essay provides an engaging narrative of Tyler's journey in basketball. The writer effectively uses dialogue, descriptions, and personal reflections to share their experiences. They show the evolution from disappointment to achievement, offering a full arc to the narrative.

Score: 25 (Excellent)

Organization and Structure (18 points)

The narrative is chronological, taking the reader through a sequence of events from initial disappointment to eventual success. The structure is mostly clear, but there are areas where transitions between events or ideas could be smoother to ensure a more seamless reading experience.

Score: 15 (Proficient)

Textual Analysis and Reference (27 points)

Tyler's personal experiences, feelings, and reactions are central to the narrative, making it authentic and relatable. The writer integrates specific game moments, interactions with friends and coaches, and personal reflections effectively. The use of dialogue brings the narrative alive, offering readers a direct insight into the moments.

Score: 24 (Excellent)

Language Use and Vocabulary (17 points)

The language is descriptive and paints a vivid picture of Tyler's basketball journey. The informal, conversational tone suits the personal narrative style. Some sentences could benefit from refinement for clarity or varied structure to enhance the narrative's flow.

Score: 14 (Proficient)

Mechanics and Conventions (10 points)

The narrative mostly adheres to grammar and punctuation rules. There are a few areas where sentence structures could be improved or where the narrative might benefit from additional punctuation for clarity.

Score: 8 (Proficient)

Final Thoughts

Tyler's essay provides a heartfelt and engaging account of his basketball journey. The writer's ability to capture moments of joy, disappointment, determination, and achievement showcases their storytelling skills. The narrative takes the reader on a roller-coaster of emotions, making it a compelling read.

Total Score: 86/100

Feedback Message:

Tyler, what a fantastic narrative! Your passion for basketball shines brightly throughout your essay, making it both heartwarming and inspiring. You've done a great job in transporting the reader to the basketball court and sharing your personal journey. Remember, every narrative is a reflection of your unique experiences, and you've captured yours beautifully. Continue to embrace both the ups and downs in your storytelling, as they paint a complete and authentic picture. Keep up the brilliant work and always shoot for the stars—both on and off the court!

Kate responded, "I think that the instant feedback would alleviate having students wait for me to get to them, which can leave them with nothing to do. It may help them become more independent and less reliant on the teacher. This would allow me more time to help students who need more support while children who have fewer needs can continue working. Essentially, we could get more done with less waiting time. How did you come up with the criteria?"

Ruthie explained, "The criteria for that writing task was based on a rubric from another state. We adjust the tool to respond to various types of writing, measured according to standards and specific rubrics. We wanted the feedback language to be more encouraging than some of the rubric language that we've seen, even though the writing advice is based on the original rubrics. Most importantly, though, we wanted language that would help the student to immediately begin the revision as the teacher circulates. No more waiting days until the teacher could get to all the papers."

Ruthie beamed with pride. Watching colleagues take an idea and run with it was always rewarding. The horizon of educational possibilities seemed to be expanding, and Ruthie was thrilled to be part of the movement propelling it forward.

CHAPTER 9

SYNERGY

Ruthie has become a WISDOMK12 evangelist; however, her enthusiasm isn't just confined to her classroom. The distinct chimes of successes in different subject areas formed a harmonious chorus in Ruthie's mind. It was this symphony of ideas that led to a sudden revelation.

After the enlightening conversations at the beginning of the new school year, Ruthie realized that the true power of WISDOMK12 lay in its synergistic potential with other progressive teaching methodologies like Writing across the Curriculum (WAC) and Project-Based Learning (PBL).

Ruthie had written an interdisciplinary unit for Pittsburgh's beloved Kennywood Park's 100th Anniversary in 1998, and the park distributed booklets to all schools in the tri-state area as an educational outreach. The unit included plans for every subject in a middle school, and the lessons included authentic activities, such as conducting polls in math and representing the data graphically, giving mini reports in science class about park topics like *How does neon glow?* or *How high can a roller coaster be built?* and suggesting themes and designing floats for Kennywood's annual Fall Fantasy Parade. She knew that the Kennywood week was everyone's favorite week of school; it was also the week where she and her colleagues saw the proverbial light bulbs going off in students' heads again and again and felt the true joy of deep learning in the classrooms. Then she thought about how ChatGPT could help teachers integrate lessons of all kinds across the curriculum.

The Transformative Power of Synergy

In our educational odyssey, we've found that the merger of WAC, PBL, and WISDOMK12 offers something extraordinary. This synergy magnifies the impact of individual teaching strategies, enhancing both student learning and teacher effectiveness.

Why does this synergy work so well?

- WAC helps students articulate complex thoughts across disciplines, honing their writing skills irrespective of the subject.
- PBL gives students real-world challenges to solve, igniting their critical thinking and collaborative skills.
- WISDOMK12 introduces the precision of AI-driven feedback and the engagement of gamification into this equation.

Together, these three elements revolutionize the educational landscape. Let's consider some examples.

INTEGRATING RESEARCH, WRITING, AND REAL-WORLD PROBLEM SOLVING

Example: Consider a science class where Ruthie collaborates with the science teacher, Tina. Students embark on a PBL task to investigate the effects of urbanization on local ecosystems.

- **WAC Component:** Students scour scientific journals, websites, and other databases to construct comprehensive reports.
- **WISDOMK12 Component:** Using WISDOMK12, they get immediate feedback on their draft reports. The AI provides suggestions for improvement, ensuring scientifically accurate and well-articulated final reports.

Collaborative Writing and Team Projects

Example: In a social studies class, students work in teams to create multimedia presentations about the cultural impacts of immigration in Pennsylvania.

- **WAC Component:** Students write specific sections, weaving in primary sources, interviews, and historical data.
- **WISDOMK12 Component:** After the initial drafts are in, WISDOMK12 assists in refining them. Gamification elements add a layer of excitement as teams can accumulate points based on the quality and depth of their work.

CROSS-DISCIPLINARY PROJECTS AND DIVERSE WRITING TASKS

Example: Imagine a joint math and English project where students explore the history and societal impacts of mathematical discoveries.

- **WAC Component:** Students write biographies of famous mathematicians and essays on the societal ramifications of their work.
- **PBL Component:** The project crescendos into a "Math Fair," an interactive exposition where students present their research.
- **WISDOMK12 Component:** As students polish their written materials, WISDOMK12 chips in with targeted feedback. Gamification adds an extra layer of enthusiasm, with badges like "Historical Expert" and "Mathematical Storyteller" up for grabs.

Conclusion

The alignment of WAC, PBL, and WISDOMK12 offers a holistic, engaging, and rigorous educational experience—the type of experiences all students will encounter and have to master in worksites across our nation It's a synergy where the sum is greater than its parts, ensuring students are not just academically successful but also life-ready. With these tools in hand, Ruthie felt more empowered than ever, looking forward to the future of her own teaching. She wasn't alone; a wave of educational transformation was just beginning to crest.

CHAPTER 10

AMPLIFYING LEARNING THROUGH THE FEEDBACK LOOP

As Ruthie and I delved deeper into the myriad capabilities of ChatGPT to enhance the educational landscape, the indispensable role of feedback in the learning process emerged as a linchpin for our vision of WISDOMK12. Feedback isn't a modern contrivance in education; it's a time-honored element that sits at the core of the teaching-learning nexus, sharpening the minds of learners while guiding the pedagogical strategies of educators.

Our exploration of feedback's role steered us through the edifying realms of Constructive Learning Theory, where feedback isn't merely an external input but also a scaffold that propels students from their current understanding to a more refined conceptual grasp. The illustrious educational theorist Vygotsky introduced us to the Zone of Proximal Development (ZPD), a domain where guided learning takes place, a space between what learners can do independently and what they can achieve with assistance. Feedback, we realized, is that crucial assistance, a beacon that illuminates the path through the fog of misunderstanding and into the clarity of comprehension.

We were inspired by how feedback could fuel metacognition, fostering a breed of learners adept at evaluating their own thinking processes and evolving through self-reflection. The catalytic effect of feedback on motivation and self-efficacy unveiled itself as a profound agent of change, forging a path for learners towards a realm of self-assured inquiry and dogged pursuit of understanding.

Unveiling the layers of benefits, we saw how feedback acts as a corrective lens, aiding in error rectification and conceptual reorientation. In a classroom bustling with diverse minds, feedback personalizes the learning experience, addressing the distinctive needs and misunderstandings of each pupil, weaving a fabric of learning tailored to individual aptitudes.

Our exploration didn't stop at the recognition of feedback's virtues but extended to envisioning its dynamism in a technology-augmented educational ecosystem. We envisioned WISDOMK12 as a crucible where the time-tested efficacy of feedback melds with the cutting-edge capabilities of ChatGPT. The promise of instant, personalized, data-driven feedback heralded a new dawn in formative assessment, providing a fertile ground for standards-based learning, where each interaction with ChatGPT moves students closer to the educational goals set before them.

The myriad perspectives of peer feedback further enriched our understanding, revealing a vista where students, propelled by collaborative analysis and collective reasoning, ascend to higher plateaus of understanding.

As Ruthie and I poised on the cusp of integrating ChatGPT's feedback mechanism within the ambit of WISDOMK12, the prospect of harnessing the Feedback Loop to personalize and amplify learning unfolded before us like never before. The ensuing chapters delineate our journey and the blueprint of engendering a robust, feedback-rich educational environment, setting the stage for an era where teaching and learning are not just about the transfer of knowledge, but a harmonious, enriching, and self-evolving journey.

The dawn of AI in education has been met with a chorus of concerns, one of the most prominent being the fear that the introduction of machines like ChatGPT into the classroom will silence the genuine, human interactions that are vital to the educational experience. The image that many have is one of students isolated, faces buried in screens, devoid of any real connection to their teachers.

Ruthie and I saw a different vision unfold in classrooms that harnessed the potential of ChatGPT properly. Far from the dystopian image that many painted, we witnessed the opposite. When used as a tool to supplement—not replace—traditional teaching methods, AI became the catalyst that amplified student-teacher communication. The feedback loop, one of the most vital mechanisms in education, took on new dimensions.

Every teacher knows the value of timely, personalized feedback. It steers students towards their goals, corrects misalignments in understanding, and fuels motivation. The traditional feedback loop, however, has its limitations. Teachers, no matter how dedicated, have constraints on their time, and it can be challenging to provide each student with detailed feedback, especially in larger classrooms.

Years ago, when I was Ruthie's student teacher, she emphasized the importance of timely feedback, ideally within 48 hours. My father died when I was in high school, so I was working my way through college as a night dispatcher at the local police station. With 166 seventh graders in her English class, I would not have been able to do it alone. Empathizing with my situation, Ruthie split the workload, but it was still difficult.

Understanding that feedback doesn't always have to come from the teacher, we passed out PQS (Praise, Question, Suggest) sheets galore, but at that grade level, it was mostly the blind leading the blind. As we later worked together as colleagues in the same district, we tried other approaches, such as training "table leaders" from among the strongest students so at least one student per cooperative group was somewhat helpful. Mostly, Ruthie kept unholy hours in order to keep the ball rolling (thus, the picture on the floor from Chapter 1).

Enter ChatGPT, equipped with the insights gained from SWIPE. The AI assists educators in sifting through student submissions, highlighting areas of strength areas in need of improvement. But this is merely the beginning of the loop.

The true power lies in how a teacher takes these insights, adds their personal touch, expertise, and understanding of the student, and then guides the student forward. The feedback loop becomes a dynamic interplay between student, teacher, and AI, ensuring that the student is always at the center, always being heard.

Ruthie once shared an instance that perfectly encapsulated this. After receiving insights from ChatGPT about a student's writing, she was able to sit down with the student and focus on the areas pinpointed by the AI. This wasn't a mechanical correction session. It was a meaningful conversation where the student could share their thoughts, express any challenges they faced, and jointly chart out a path forward with Ruthie. The AI played a crucial role, but the human touch was irreplaceable.

This is the future we envision—where technology and human connection coalesce, where the feedback loop is tighter, more insightful, and ever-evolving. As we navigate through this chapter, we'll uncover the intricacies of this enriched feedback loop, exploring how it can transform teaching and learning, making them more dynamic, adaptive, and student-centric.

RUTHIE'S CHATGPT FEEDBACK LOOP

Mid-semester had arrived, and Ruthie's focus turned towards the state standards for argumentative writing. She assigned her students a task:

Some people think that physical education should no longer be a graduation requirement. Individual school districts may choose to offer the program or to eliminate it as a requirement. Write to your school board to persuade it to maintain or to eliminate the current program.

Among the pile of essays, one stood out. It was from Kent, a student who had always exhibited potential but often grappled with articulating his thoughts in a structured manner.

This was Kent's essay:

Some people think that physical education is not necessary for school. Some people think that physical education is necessary. Today I will be talking about my opinion on physical education and if I believe it should be taught, or shouldn't.

I believe that physical education should be taught because it gives students a time to get away from the pens, pencils, chromebooks, and the teaching environment, and a time to show your athletic ability, get a good workout in, and most importantly have fun! Gym is also a time to talk with friends and fool around a little bit! I will take you on a journey of me being in the gym and having a great time!

I woke up this morning eager to get to school because we had gym first period! Gym is my favorite period of the day because of all my friends in class! After I made my cereal, took a shower and brushed my teeth I hopped on the bus and was on my way to school. When I got to school me and my friends started picking team captains for floor hockey. My friend Lucas and I got picked as team captains. We played rock paper scissors for first pick in the drafts. I ended up winning rock paper scissors my 2 wins. The first person I picked was Jack. Jack and I played hockey together for 3 years. We were a great team. After, Lucas picked Jaber because he was funny. We started our game. The game went on goal after goal after goal for my team Lucas's team with zero points. After the game we all changed back into uniforms and talked about the game while Jack and Jaber played catch. Then we headed back to class to get ready for work.

And that's proof of great memories you can make at gym.

Wanting to provide detailed feedback, Ruthie ran Kent's

essay through ChatGPT, setting it against her crafted rubric aligned with state standards for argumentative writing. In no time, ChatGPT delivered its analysis.

Rubric Grading:

1. **Focus (15/20)** The student does maintain a clear focus on the importance of physical education, education, reflecting personal enthusiasm and a positive experience in gym class. However, there's a slight deviation from the initial purpose of discussing whether it should be a part of the school curriculum, as the narrative mainly becomes a recount of a specific day in gym class rather than providing broad arguments.
2. **Content (13/20):** The student provides a personal anecdote to illustrate the benefits of physical education, which makes the content engaging. However, the essay lacks diverse points and evidence that support the necessity of physical education beyond the individual experience. Additional perspectives or statistics on the benefits of physical education could enhance the argument.
3. **Organization (14/20):** The narrative has a clear beginning, middle, and end, which makes for a structured recounting of the gym class experience. However, the organization of the argument in favor of physical education is not as cohesive, as there could be a more thorough introduction, a clearer development of the argument, and a more succinct conclusion that restates the main points.
4. **Style (14/20):** The student's voice is casual and friendly,

which suits the narrative style of the piece. The use of direct address ("I will take you on a journey") adds a personal touch. Nonetheless, the style could be improved by varying sentence structure for effect and by including more persuasive language to strengthen the argument.

5. **Convention of the English Language (14/20):** While the essay is mostly free of errors in spelling, there are issues with punctuation and grammar that impact readability. Additionally, the transitions between recounting personal experience and arguing the point could be smoother to improve the overall flow of the text.

Recalling the SWIPE entries Kent had submitted earlier in the year, Ruthie was better prepared for their feedback session. She seamlessly integrated the talking points into their conversation, ensuring the feedback was not only constructive but also deeply personalized.

"Kent," Ruthie began, referencing the talking points, "your writings have consistently shown a sincere attempt to provide details to support your argument. Remember your school uniform essay? Just like that, this essay also exhibits multiple vibrant ideas, but some of them do not support your central argument. Can you identify the central argument?"

Kent searched for a while and then pointed to the text as he read. "I believe that physical education should be taught because it gives students a time to get away from the pens, pencils, chromebooks, and the teaching environment, and a time to show your athletic ability, get a good workout in, and most importantly have fun!"

"Good," continued Ruthie. "That can be refined a bit, but which sentences in this essay do not support or contain evidence for that central idea?"

After rereading, Kent asked, "Is it the sentence about taking a shower and brushing my teeth?"

"That's one of them, but there's more. Let's read the feedback again."

After reviewing the feedback together, Kent suggested, "Take out the rock, paper, scissors line."

"That's part of it, too, but there's still more than doesn't support the central argument."

"Do you mean the details about the gym class?" asked Kelly. "I was providing evidence of how fun gym class can be."

"Yes, Kent, but that was your individual experience on a given day. It doesn't serve as evidence in all situations. You briefly mentioned a break from the teaching environment, an opportunity to show your athletic ability, and to get a good workout in. Can you think of other benefits of physical exercise in general?"

"You always tell us that physical activity improves your brain," he replied.

"Yes! And how about benefits like goal setting, maintaining self-discipline, getting along with peers, and developing a sense of teamwork and good sportsmanship? Exercise also burns calories and helps people to sleep better."

"Oh, yeah. Those are all good."

"For your revision, I'm going to ask you to do research to support your argument more fully. For the first draft, I wanted everyone to think on their own as they would have to on a standardized test. We're still learning to write persuasively, but now we'll look for strong arguments to support your position."

"I can do that!" said Kent.

"I know you can," Ruthie assured him.

With the synthesized insights from SWIPE entries and the actionable talking points, Ruthie and Kent co-developed a roadmap for Kent's writing improvement. It was evident: the amalgamation of AI's precision and Ruthie's human touch had transformed the feedback session into a comprehensive learning experience for Kent. The loop was not just about corrections, but about genuine understanding and growth.

BUILDING A FEEDBACK LOOP CULTURE IN YOUR CLASSROOM

The success of Ruthie's session with Kent showcased the potential for every student to experience transformative feedback. The combination of personal attention and ChatGPT's precision not only illuminated areas of improvement but also affirmed strengths. The challenge then arose: How to replicate this depth of interaction with every student, given the constraints of time?

The solution lies in the systematic application of ChatGPT. Here's a step-by-step guide for teachers:

- **Initiate with SWIPEs:** Always begin with students' SWIPE entries. These profiles form the bedrock, offering a deep understanding of each student's personality, aptitudes, and areas of development.
- **Rubric-Based Analysis:** Use a standard rubric that

aligns with your curriculum's benchmarks. Feed student work and the rubric into ChatGPT. The AI will assess the work against the rubric, providing in-depth feedback on each element.

- **Bulleted Talking Points:** Upon analysis completion, ChatGPT will craft a concise list of talking points based on the rubric results and the SWIPE profile. This ensures that conversations are direct, addressing the most crucial aspects of the work.
- **Frequent Micro-Sessions:** Given that ChatGPT delivers feedback so efficiently, teachers can now engage in regular, yet brief feedback sessions with students. These "super-sized" micro-sessions, lasting just a minute or two, are hyper-focused. With ChatGPT's support, teachers can zero in on precise areas of need or commendation without the need for prolonged meetings. And since ChatGPT retains all previous interactions and feedback, every new session builds upon the last, ensuring continuous progression.
- **Personal Touch:** The heart of the feedback process remains the teacher's empathy, context, and human touch. Always start with acknowledging positives, highlight genuine progress, and collaboratively chart the next steps with the student.
- **Iterate and Improve:** After each feedback micro-session, take a moment to reflect. What went well? What could be enhanced? This reflection, over time,

fine-tunes the procedure, promising an optimized experience for every student.

Through this structured approach, not only do teachers have the tools to provide timely and impactful feedback, but they also foster a consistent feedback loop. It becomes a dynamic blend of technology and human connection, each amplifying the other to create an environment ripe for student growth.

CHAPTER 11

THE POWER OF REFLECTION

Progressive education reformer John Dewey (1859–1952) asserted, "We do not learn from experience. We learn from reflecting on experience." Yet for decades (and probably throughout your own primary and secondary education), there was little or no reflection going on in classrooms. Students looked at the grade (which was often arbitrary before rubrics) and threw the assessment away.

The people who were applying Dewey's philosophy all along were not the teachers, but rather the coaches. Typically, a football or basketball team would play on a Friday night and gather back at school on Saturday morning to review the film and plan from there. Sports teams have always done what students need to do—prepare, execute, reflect, and revise: What does each player need to do individually? What do we need to improve as a team? What new skills do we need to learn? What do we need to review? What will we need to do differently for the next opponent?

Ruthie's and my classrooms used portfolio assessment to foster reflection, and that is still our recommendation. What ChatGPT does so well, though, is to provide timely, specific feedback as a basis for reflection and subsequent revision. It opens the door for frequent reflection and rapid improvement.

As the feedback loop with ChatGPT becomes a staple in the classroom, a significant transformation in teacher-student interaction begins to unfold. Ruthie and I envisioned a future where these brief yet potent feedback micro-sessions become the heartbeat of the learning process.

One cannot overlook the emotional impact of positive feedback conversations. That kind of feedback fires up entirely different regions of the brain, and the more regions that are activated during the learning process, the more likely new information will "stick."

In traditional settings, in-depth teacher-student conversations were often confined to sporadic parent-teacher meetings, rare one-on-one sessions, or fleeting moments grabbed in busy classrooms. However, with the implementation of a feedback loop culture, the very fabric of this interaction changes. Instead of a few isolated in-depth sessions, teachers and students could now potentially engage in hundreds of these "micro-interactions" over an academic year. This leads to an explosion in the sheer volume of meaningful teacher-student dialogue, potentially multiplying the interaction time manifold.

Yet, it's not just about quantity; it's about the quality and continuity of these interactions. Repeated exposures over time

= retention. By both the teacher and student entering a brief reflection after their micro-session, they further solidify the feedback process. This simple act of reflection does wonders:

1. **Deepens Understanding:** Taking a moment to reflect allows both parties to internalize the feedback and insights from the conversation. This introspection bridges any gaps in understanding and ensures alignment in objectives.
2. **Guides Future Interactions:** These reflections act as a compass, guiding future discussions. Over time, ChatGPT, by analyzing these reflections, can tailor its insights even more, ensuring that each subsequent conversation is sharper, more relevant, and more impactful than the last.
3. **Builds Trust and Rapport:** When both teacher and student are actively involved in the reflection process, it fosters mutual respect and understanding. It shows that both parties are deeply invested in the learning journey, creating a strong bond of trust.
4. **Fosters a Culture of Continuous Improvement:** This iterative process, where every interaction is followed by reflection, breeds a culture of continuous improvement. It's a cycle where feedback spurs growth, which in turn leads to more refined feedback.

In essence, the feedback loop culture, amplified by ChatGPT, doesn't just maintain the essence of teacher-student communication; it elevates it. This approach promises not just a transactional exchange of information, but a transformational journey of understanding. It paves the way for a future where students and teachers, in tandem with technology, craft a seamless tapestry of communication, understanding, and growth.

FEEDBACK IS A CORNERSTONE OF EFFECTIVE TEACHING AND LEARNING.

Here's a synthesis of the scientific reasoning behind feedback's critical role:

1. CONSTRUCTIVE LEARNING THEORY:

- Feedback supports Constructivist Learning Theory where learning is seen as a process of building upon prior knowledge. By receiving feedback, students can see where their understanding is correct or incorrect, allowing them to adjust and refine their knowledge.

2. ZONE OF PROXIMAL DEVELOPMENT (ZPD):

- The concept of the Zone of Proximal Development, introduced by Vygotsky, refers to the difference between what a learner can do without help and what they can do with guidance. Feedback acts as that guidance, helping students navigate through their ZPD and reach higher levels of understanding.

3. METACOGNITION ENHANCEMENT:

- Feedback helps enhance metacognition, the ability to think about one's own thinking. It encourages students to reflect on their learning processes and understand what strategies work best for them.

4. MOTIVATION AND SELF-EFFICACY:

- Timely and constructive feedback can boost student motivation and self-efficacy (a belief in one's ability to accomplish tasks). It provides students with a sense of achievement and a clear path to improvement.

5. ERROR CORRECTION AND CONCEPTUAL CHANGE:

- Feedback helps in error correction and addressing misconceptions. It assists students in moving from incorrect to correct understanding, thereby facilitating conceptual change.

6. FORMATIVE ASSESSMENT:

- Formative feedback, delivered during the learning process rather than at the end, provides students with valuable information that can help shape their learning as it's occurring. It's a vital component of formative assessment.

7. PERSONALIZATION OF LEARNING:

- Feedback can personalize learning by addressing the unique needs and misunderstandings of each student, thereby making learning more effective and tailored to individual needs.

8. STANDARDS-BASED LEARNING:

- Feedback that is aligned with learning standards helps students understand where they are in relation to expected learning outcomes and what steps they need to take to meet or exceed those expectations.

9. RETENTION AND TRANSFER:

- Feedback can aid in the retention of learned material and the transfer of learning to new contexts by reinforcing correct understanding and providing corrective measures for misconceptions.

10. PEER LEARNING:

- Peer feedback fosters a collaborative learning environment where students can learn from each other's perspectives, promoting a deeper understanding of the subject matter.

In an era of technology integration in education, AI-enhanced feedback mechanisms like those provided by ChatGPT can take these benefits to new heights by offering instant, personalized, and data-driven feedback, while freeing up teachers' time for more in-depth, personalized interaction.

*****

Ruthie's epiphany about the potential of an interdisciplinary project-based learning competition was rooted in a deeper understanding of how learning is changing in the 21st century. Gone are the days when students were merely receptacles for knowledge; today's learners need to be agile thinkers, capable of connecting the dots across disciplines and applying their knowledge to real-world scenarios. This shift in pedagogical focus aligns perfectly with the capabilities of ChatGPT.

Contextual Learning

At the heart of Project-Based Learning (PBL) is the idea of providing students with real-world contexts for their learning. When students are tasked with solving a genuine problem or answering a complex question, they are more motivated and engaged. ChatGPT, with its vast knowledge base, can provide information, context, and nuanced answers to students' queries, supporting them as they navigate through their projects.

Collaboration and Communication

PBL emphasizes teamwork. Students often work in groups, sharing ideas, debating solutions, and dividing tasks based on individual strengths. ChatGPT can facilitate this by helping students articulate their thoughts, offering suggestions for collaborative tools, and even aiding in the creation of presentations or documents.

Critical Thinking and Research

Research is an integral part of PBL. As students delve into their projects, they're required to sift through vast amounts of information, discerning what's relevant and credible. ChatGPT can guide students through this process, helping them identify reliable sources, offering summaries of complex topics, and challenging them to think critically about the information they encounter.

Reflective Feedback

Reflection is key to PBL. Students need to evaluate their progress, understand where they might have gone wrong, and determine how to improve. ChatGPT can aid in this reflective process, providing feedback on students' ideas, offering suggestions for improvement, and prompting students to think deeply about their decisions.

Cross-Disciplinary Connection

One of the most exciting aspects of PBL is its potential to bridge subjects. For example, a project might require students to use their math skills to solve an historical problem or their scientific knowledge to address an issue in literature. ChatGPT, with its interdisciplinary knowledge, is perfectly poised to assist students in making these cross-curricular connections, ensuring they see the bigger picture and understand the interconnectedness of their learning.

For our team, the benefits were clear. Using ChatGPT in conjunction with PBL was like giving students a super-powered assistant, always ready to help, guide, and challenge them as they embarked on their learning journeys. As we began to think of ways weave this dynamic duo into curricula, the possibilities seemed endless.

THE PROJECT: "SCHOOL OF THE FUTURE: CRAFTING AI-DRIVEN EDUCATION"

Ruthie's imagination swirled with the idea of students thinking forward, and she was adamant about integrating technology into their vision. "School of the Future: Crafting AI-Driven Education" was born out of this idea. The project would task students to imagine a school in the year 2050 where AI like ChatGPT was a staple in enhancing the learning experience.

The goal? To design an educational institution that was adaptive, personalized, and maximized the potential of every student. By doing so, the students would also realize the potential that AI has in their education right now.

English/Language Arts:

Students would be required to craft narratives or scripts that depicted a day in the life of a student in their AI-driven school. How would storytelling change when AI could craft narratives or when books could be interactive? ChatGPT would assist in refining their writing, suggesting areas of improvement, and even give them a taste of AI-generated storytelling.

History/Social Studies

The students would look back at the evolution of education, from one-room schoolhouses to virtual learning during pandemics, to project the trajectory of education. With ChatGPT, they could draw parallels between past educational shifts and their proposed AI-driven changes, understanding the societal impacts of these shifts.

Science

Students would delve into the science of AI. How does machine learning work? What's the potential for bio-integrated technology in learning? Using ChatGPT, they'd gain insights into the latest AI research, understanding both the potential and the ethical implications of such advancements in education.

Math

With the challenge of creating a budget for their futuristic school, students would project costs, return on investment (ROI) of AI educational tools, and consider economic scalability. ChatGPT would be an ally in their mathematical endeavors, aiding in complex calculations and providing data trends in tech education investments.

Art

Visualization would be key. Students would create models, digital designs, or virtual reality tours of their envisioned AI-driven school. ChatGPT could provide insights into digital design techniques, historical and futuristic design references, and even AI-generated art.

Scoring with ChatGPT

With a holistic rubric in hand, the emphasis would be on innovation, feasibility, interdisciplinary integration, and overall vision clarity. We began to discuss how ChatGPT could impartially evaluate each segment of the project against the rubric, ensuring fairness in the competition while capturing the essence of each student's vision for the future of education.

Anxious to share her vision with the WISDOMK12 team, Ruthie felt a new sense of purpose. If successful, this competition could be the academic highlight of the year in a school, community, or even a diocese or county.

CHAPTER 12:

RUTHIE'S NEW IMAGINARY HIGHLY EFFECTIVE TEACHER FRIEND

After a long day at headquarters where we hand scored student essays and compared them to ChatGPT's feedback, we began to tweak the language of the AI feedback. Early feedback mimicked the rubrics themselves, but phrases like "Generic use of a variety of words and sentence structures that may or may not create writer's voice and tone appropriate to audience" were not only unclear, but also discouraging when found in the "Proficient" column. We were working toward encouraging, growth-mindset feedback while staying true to the skill levels represented by the numbers and descriptors. We also wanted to ensure that the language was clear and understandable to students at various grade levels.

Ruthie and I stopped at a local restaurant that night to debrief over dinner, and there we had a conversation that affirmed all we had been doing, not just this year, but since we first met when I was a student in her class decades earlier.

"Mike, we've been through so many educational initiatives together over the years, but integrating ChatGPT is not just changing a rule—it's changing the entire game.

"Let me tell you a story. When calculators first came on the market, my ex-husband (also a teacher) and his siblings pitched in to buy one for their father, a civil engineer. The calculator was not a scientific calculator; it was the kind you can buy at the drug store now for $4.99. In 1975, that calculator cost $125.00. We borrowed the calculator at grade time, and our fellow teachers waited in line to use it. Before that, we had to figure each student's grades the old-fashioned way, adding and dividing across each student's row of grades. "Weighing" quizzes, tests, homework assignments, and projects complicated the calculations further, so many teachers stuck to a consistent 100-point scale. These same teachers who were thrilled about using the calculator at grade time, however, bemoaned the invention, fearing that students would never learn math if they knew they could "just use a calculator." Calculators were forbidden in school, and at Open House, teachers warned parents against allowing children to use them at home. That's exactly where educators are right now with AI and apps like ChatGPT."

"Right!" I agreed. "I just met with a group of teachers who were resistant. From what they had heard in the media over the past months, they essentially said, 'Thanks, but no thanks. There's a committee that's going to develop a policy. We'll wait and see what they say."

"I'm not at all willing to wait. I've been telling my students about the ways I use ChatGPT, and we'll use it together when it's ethical and appropriate. Their aversion to writing seemed to vanish when they knew they were going to get AI feedback in addition to my own on the pilot writings."

When Ruthie worked with teachers as an ELA Facilitator for Johns Hopkins University Talent Development Secondary, their training in Baltimore largely consisted of Dr. Mariale Hardiman's Brain-Targeted Teaching Framework. With that background, facilitators could assist teachers by analyzing their lessons and methods in terms of how the brain learns. "We were told in 2015 that brain scans showed that the then five-year-old's brains were wired differently from any generation before due to the daily use of electronic devices. Dr. Hardiman warned that we would have to adjust our teaching significantly in order to reach them. I didn't understand that at the time, but these past few years, I see the difference. Those five-year-olds are the middle schoolers I teach now. The new challenges that teachers face—from disinterest to obsession with electronics, students' tendency to shortcut everything, short attention spans, and the erosion of grammar rules—are often lumped under 'pandemic loss,' but I think it has more to do with the digital age in society."

"So what's the answer?" I asked.

"All I know is that starting this year with ChatGPT has given me renewed energy and hope for the future. Last spring when you and I talked about how I caught my students cheating by having ChatGPT compose their final writings, I was at a loss of where to go from there. All the writing assignments for the last month of school took place in front of me in class rather than for homework. Now ChatGPT is going to assist me in speeding up the writing process. It saves me time in planning. I'm using it in multiple ways and feel like I'm on the frontlines in education in the age of AI and moving forward."

"Me, too," I concurred. "I feel like all we've done previously as educators led us to this day. It's fun, isn't it? This is what plumbers must've felt like when they introduced indoor plumbing. By the way, I saw your sister Gigi at the store, and she told me to ask you about Channa. Who's Channa?"

Ruthie laughed, then covered her face to show embarrassment. "It's my name for ChatGPT, my new imaginary highly effective teacher friend. Channa is the name I would have given to a daughter, but I had three sons. It also uses the first three letters of ChatGPT, so it felt like the right name. I feel a little foolish explaining that, but it really does feel like I have an extremely competent personal assistant."

"So what did Gigi think you were going to tell me? We've been doing this together."

"When Gigi first got Alexa years ago, she would ask it things all day and then tell me about what Alexa said. I said it was like she had a roommate. Now she says I've topped her by far because I can't stop talking about Channa, my new imaginary highly effective teacher friend. I'm always asking ChatGPT things like, 'Generate Christmas door decorating ideas,' or 'Suggest non-tangible rewards that middle schoolers would like.' Lately, though, I've been asking deep philosophical questions that help me to reflect on my own practice. Even though I presented a session at the Diplomas Now national conference called "Vocabulary: The Key to Academic Success," I asked Channa to describe best practices in teaching vocabulary. (I was right on target but still gained an idea.) Asking Channa is way beyond asking Google because of the way ChatGPT synthesizes information. I can't stop playing with it."

"You really are a convert, aren't you?"

"Yes! The students are overtly excited when we use AI, and so am I."

CHAPTER 13

EMBRACING THE FUTURE, TOGETHER

As we end our journey with Ruthie, ChatGPT, and the transformative power of AI in education, it's essential to remember that we're just at the dawn of this new era. What Ruthie and I have shared throughout these pages is merely a glimpse of the possibilities that lie ahead. Yet it is a truly significant glimpse, one that offers a compelling vision of what education can be when we harness technology's potential for the greater good.

Change, especially in a domain as personal and entrenched as education, can be daunting. Yet, it's often the bold leaps, the willingness to embrace innovation, that bring about the most profound and lasting impacts. Ruthie's journey wasn't one of instantaneous success; it was filled with curiosity, trial, adaptation, and growth. She took risks, sought out knowledge, and continuously reflected on her practice. But through it all, her focus remained steadfast: improving learning outcomes for her students.

Now we extend that challenge to you. We invite you to not just read about these tools and strategies but to put them into practice. Try out the prompts, experiment with the feedback loops, delve into PBL, and see firsthand how these tools can reshape your teaching landscape. And as you do, remember that you're not alone.

The beauty of the digital age is that it has brought educators closer together than ever before. Networks of passionate teachers are forming, eager to share their successes, challenges, insights, and questions. Join this vibrant community. Connect, collaborate, and contribute to this ever-evolving conversation.

We urge you to remember the importance of connection in all this. While AI and technology play a crucial role in this story, the heart of education will always be the relationships we build: teacher to student, student to student, and yes, teacher to teacher.

So, as you turn this final page, remember that this isn't the end, but merely the beginning. Take the leap, share your journey, and let's shape the future of education, together.

Here's to innovation, exploration, and endless learning. Join the conversation, and let's craft the next chapter of education, hand in hand.

INSTANT EXPERT PROJECT

Introduction

Welcome to the "Instant Expert" challenge! Your mission, should you choose to accept it, is to dig deeply into a topic of your choice, and with the help of research tools and perhaps even AI assistance, emerge as our class's "Instant Expert." By the end of this project, you'll be ready to share your newfound expertise with all of us. Ready? Let's dive in!

Objectives

By the end of this project, you will:

- have developed sharper research skills.
- be more confident in your presentation abilities.
- possess more in-depth knowledge about your chosen topic.
- have honed your critical thinking and analytical skills.

Instructions

1. Selecting Your Topic

Choose from the provided list or propose a unique topic to the teacher for approval.

2. Research

Use a combination of textbooks, online resources, and AI tools (like ChatGPT) to research your topic. Remember to note your sources!

3. Prepare Your Presentation

This can be in the form of slides, a short video, a poster, or any other creative format. Ensure your presentation covers:

- key concepts or elements of your topic.
- interesting facts or discoveries you've made during your research.
- any challenges or opposing views related to the topic.
- your personal reflections or takeaways.

4. Presentation Time

You'll have 5-7 minutes to present. Be ready to answer questions at the end.

Evaluation

Your project will be evaluated based on

- depth and accuracy of research.
- alacrity and creativity of your presentation.
- ability to engage your audience and answer questions.

Tips for Success

- Start early! Give yourself plenty of time for research.
- Practice your presentation to ensure you stay within the time limitations.
- Be confident! By the end of your research, you will be the class expert on this topic.

Remember, the aim is not just to learn but also to enjoy the process of discovery. Embrace your inner expert and have fun with this project! We can't wait to learn from you.

Appendix A: Instant Expert Topic Suggestions

The following list offers a curated selection of "Instant Expert" topics across various subjects. These are designed to inspire and guide educators in providing rich, interactive learning experiences using AI tools like ChatGPT. Dive deeply into these subjects with your students and explore the vast knowledge within each domain.

English Literature:

Shakespearean Sonnets: Their Structure and Significance
The Gothic Novel: Characteristics and Key Works
The Harlem Renaissance: A Literary and Cultural Movement
Modernist Literature: Breaking Traditional Molds
Feminism in Literature: From Early to Contemporary Works

Algebra:

Quadratic Equations: Understanding Roots and Vertex
Systems of Equations: Graphical and Algebraic Solutions
Polynomials: Degree, Terms, and Factoring Methods
Linear Functions: Slope, Y-intercept, and Graphing
The Binomial Theorem: Expansion and Applications

Biology:

Photosynthesis: How Plants Convert Light to Food
Cellular Respiration: The Process and Its Stages
Genetics: Understanding DNA, Genes, and Heredity
Evolution: Natural Selection and Adaptation
The Human Digestive System: From Ingestion to Absorption

US History:

The American Revolution: Causes and Key Battles
Civil Rights Movement: Key Figures and Turning Points
The Great Depression: Causes, Impact, and Recovery
The Louisiana Purchase: Expansion and Exploration
The Cold War: Tensions between the US and USSR

Art:

The Renaissance: Artistic Innovations and Key Figures
Cubism: Understanding Picasso and Braque
Pop Art: Warhol, Lichtenstein, and Cultural Commentary
Abstract Expressionism: Movement and Meaning
Art Nouveau: Origins, Characteristics, and Influence

Music:

Classical Music: Key Composers from Bach to Beethoven
Jazz: Its Origins and Evolution
Rock 'n' Roll: From Elvis to the British Invasion
Hip-Hop: Roots, Evolution, and Impact
World Music: Exploring Different Cultures through Sound

Physical Education:

Anatomy of a Workout: Warming Up, Training, and Cooling Down
The Importance of Flexibility: Stretching and Its Benefits
Sports Nutrition: Fueling for Performance
Basic Biomechanics: How Movement Works
Team Sports vs. Individual Sports: Skills and Benefits

Career Technical Education:

Basic Principles of Engineering: Design and Problem Solving
Digital Literacy: Navigating Today's Technological Landscape
Entrepreneurship: Starting and Running a Business
Green Technologies: Innovations for a Sustainable Future
Culinary Arts: Techniques, Cuisine Types, and Presentation

GLOSSARY

AI (Artificial Intelligence): A field of computer science that aims to create machines capable of intelligent behavior.
Algorithm: A set of rules or procedures that a computer follows to perform a task.
API (Application Programming Interface): A set of rules and protocols that allow different software programs to communicate with each other.
Bias: In AI, refers to systems that systematically and unfairly discriminate against certain individuals or groups.
Big Data: Extremely large data sets that can be analyzed to reveal patterns, trends, and associations.
Chatbot: A software program designed to simulate conversation with human users.
Classification: In machine learning, the process of sorting data into predefined categories.
Cloud Computing: The delivery of various services over the Internet, including storage and processing power.

Conversational Interface: A user interface that mimics human conversation.

Data Analytics: The process of examining data to uncover hidden patterns, correlations, or other insights.

Data Mining: The process of discovering patterns in large data sets through the use of machine learning.

Data Set: A collection of data used for analysis.

Deep Learning: A subset of machine learning that involves neural networks with multiple layers.

GPT (Generative Pre-trained Transformer): A machine learning model for natural language understanding and generation.

Machine Learning: A subset of AI that enables computers to learn from data and improve over time.

Model Training: The process of teaching a machine learning model to improve its predictions or decisions based on data.

Natural Language Processing (NLP): A field of AI that enables computers to understand, interpret, and respond to human language.

Neural Network: A computational model inspired by the human brain, used in machine learning.

Open-Source: Software for which the original source code is made available to anyone.

Personalization: The use of algorithms to deliver personalized content or services.

PBL (Project-Based Learning): An instructional method where students learn by actively engaging in real-world and personally meaningful projects.

Prompt: A question or statement that instructs the model on what to generate in its response.

Rubric: A guide for grading or scoring academic work, often based on a set of criteria.

SWIPE (Student Writing & Interest Profile Entries): A tool for assessing different writing skills and personalizing learning.

Supervised Learning: A type of machine learning where the model is trained on a labeled dataset.

Token: In NLP, a unit of text that the model reads in one chunk.

Unsupervised Learning: A type of machine learning where the model is trained on an unlabeled dataset.

User Interface (UI): The space where interactions between humans and machines occur.

Validation Data: Data used to test the performance of a machine learning model.

Writing across the Curriculum (WAC): an approach to learning designed to encourage educators across all content areas to use writing as a tool for thinking.

WisdomK12: An online tool designed for educators to harness the power of ChatGPT in the classroom.

By harnessing the power of AI, WISDOMK12 optimizes teaching and learning in ways that supercharge and delight both teachers and students.